SOLVING PIPING NETWORKS WITH YOUR PC
By DH Berry, PE

© 2014 by DH Berry, PE
No part of this document may be reproduced, stored in a retrieval system, or transmitted in any form or by any means, electronic, mechanical, photocopying, recording, or otherwise without the prior written permission of the author.

02/18/14

00 FORWARD

SOLVING PIPING NETWORKS WITH YOUR PC
by DH Berry, PE

Welcome to my 1st ever book. Bear with me, as I am a newbie author. I hope to spark as much interest in the subject for the reader as I have.

First of all, let me make clear that this is NOT a program. It is how to USE TK Solver or other programs to solve a type of common engineering problems. TK Solver is a very powerful math program and is well suited for the type of problems outlined (and solved) in this small book. Note that this book is not trying to compete with the big (and costly) commercial programs, but is geared more toward the Mechanical engineer faced with routine network problems. It will solve problems that a lot of engineers run into, such as fire sprinklers and pumps in piping systems.

Note that I am not associated with TK Solver in any way. In fact, what I used for this book is merely the Student Version. The full blown version is more powerful. But, I have not even used the full potential of the Student Version. This is not a commercial pitch for TK. You can use any software that can solve systems of equations.

How did this book come about? I was trying to solve some simple network problems and soon realized that I was facing several non-linear simultaneous equations. That term is kind of a joke. Most mathematical descriptions of physical laws are non-linear. I had TK Solver and decided to give it a try. I read up on it and discovered it can handle something like 43000 simultaneous equations!! That should be enough. Also, the equations can be in any order, as long as all unknowns are defined AND that for n unknowns there are n equations. You probably already know that, but it is very important to avoid frustrations as I have run into.

This book is for those somewhat familiar with fluid flow and mathematical descriptions of fluid flow problems, such as Mechanical, Civil and Chemical Engineers. If not, you may have a bit of a problem. The problems outlined and solved are real-world type problems.

The book is also dedicated to my wife for her great patience since it took up a lot of my "spare" time.

Bring your notepad, pencil, PC and imagination and let's solve some network problems.

TABLE OF CONTENTS

00 FORWARD AND TABLE OF CONTENTS
CHAPTER 01 - INTRODUCTION
CHAPTER 02 – TK SOLVER AND OTHERS
CHAPTER 03 - BASIC TRAINING
CHAPTER 04 – METHODOLOGY
CHAPTER 05 – SIMPLE NETWORKS
CHAPTER 06 – SOME LOOPS
CHAPTER 07 – DIVIDED FLOW
CHAPTER 08 – FIRE PROTECTION
CHAPTER 09 – PUMPS IN NETWORKS
CHAPTER 10 – AIR DUCTS
CHAPTER 11 – FANS IN NETWORKS
CHAPTER 12 – COMPRESSED AIR NETWORKS
CHAPTER 13 – NATURAL GAS NETWORKS

CHAPTER 01 INTRODUCTION

Piping networks show up all over the place. Examples are potable water systems, sanitary and storm drain systems, fire protection systems, water distribution, chilled and hot water hydronic systems, natural gas distribution and compressed air. There are many more of course. All have one thing in common in that all are interconnected pipe. All have one or more sources of flow and one or more outlets. They can also have pumps or reservoirs. The most intricate network is of course your blood circulatory system. It is a closed system (unless you bleed) and has 1 very reliable pump. It is a positive displacement type of pump by the way.

Note that these are practical problems, not merely theoretical school homework problems.

Using the techniques in this text, and TK Solver, the reader will be able to solve moderately complex network problems, such as water supply, drains, pumped systems and fire sprinkler design.

Yes, there are more powerful programs available. Some are quite expensive, and some are free. For instance EPANET will solve water networks, and it is **Free**. I have it and have tried it out and became frustrated with it. Yes, you can model 1 or more pumps but you do not know how accurate it models them. I also find the graphical interface a bit clumsy. Also, it is only for water. It cannot handle compressed air, natural gas, or any other fluid. This text will show you how you can model most any Newtonian fluid. That is, all liquid or all gas (considered incompressible) and no plastics or slurries. Plus, (a BIG plus) you are in control of the equations. In fact, you have to be because the techniques shown require you to think and formulate the problem in a way that TK Solver or other programs can handle. The equations will NOT write themselves.

Note that the techniques are for moderately complex systems. It is probably not the thing to use to model the water distribution system for Chicago. The author has found that it works best if you are sure of the direction of flow. If you guess wrong, you get an error (as in trying to calculate LN of a negative number or division by 0) or exponential problem with Hazen Williams. If you are pretty sure of the direction of flow, the size of the system is limited by your patience in writing the equations. Systems with more than 2 loops seem to be a problem because it becomes harder and harder to determine the direction of flow in the loops.

CHAPTER 02 TK SOLVER AND OTHERS

This is a program I would have loved to have when I was a student. It is a real marvel. You can input equations in any order and as long as there are n independent equations for n unknowns, it will solve them. For hundreds of years, mathematicians and engineers have developed solutions to problems but they were not practical because of the enormous amount of calculations required. Now with computers, we can solve them. Note that you cannot cheat with dependent equations. Each equation must be independent, otherwise, no unique solution.

Once you have all equations in place, and input guesses for the unknowns (usually 1 is sufficient as a guess), TK Solver will find the solution incredibly fast. Even on a relatively slow laptop, it seems instantaneous (a fraction of a second). It is recommended to save the solution as a PDF for printout later. You should edit the solution report, because it seems that every other page does not have useful information. If you want the comments in the report, print in Landscape.

One of the peculiarities of TK is that sometimes, it hangs up when it says "dependency error". The only way out of it seems to be to close out TK and start over. This is another very good reason for saving your input BEFORE solving it.

During the course of finding solutions to various problems, timesavers and shortcuts were developed that take advantage of TK Solver's features. These will be evident as we move along.

Another program that is good to use, and in some ways easier is **Engineering Suite 1.04.** It is **FREE** and can be downloaded here: https://code.google.com/p/engineeringsuite/
The program information says the only limitation to the number of equations is the RAM capacity of your computer. It has some neat features such as rendering your equations just as you would write them which helps in troubleshooting, and you input your guess for variables just once. Also you can input the initial guess for an unknown just once, and add comments along the way. Unfortunately, the rendered equations cannot be copied or printed. Just a quirk in the program. You do not have to save first to try another solution either, because it will remember your inputs. When a solution is complete, the LOG file will show the input data. The input file can be easily saved also. Final results can be easily saved as a PFD or text file that will show up later. You can change the methods of solutions too (4 available), and it will solve problems that TK has difficulty with. Then again, TK can solve some that Esuite cannot. Get familiar with both, as it is always good to have a backup. Both Esuite and TK have the handy feature of being able to copy whole lines. This is very useful for writing the equations for pipe factors. You copy the first line for as many pipes as in the problem, then change the constants like L1 and D1 to L2 and D2. This will become evident as we go along.

Other programs that could work are MATLAB and MAPLE, though I have no experience with those programs

CHAPTER 04 - METHODOLOGY

Fluid flow depends on pressure drop. With no pressure drop or pressure difference, there is no flow. This is basic and the equations are built upon that. Also, conservation of mass dictates that what goes into the piping must come out. It seems trivial but it is also what the equations are built on.

Mathematically, pressure drop = dp = K1 x Q1^2 for a pipe. With K1 = 0.03108 x f1 x L1/(d1^5).

K1 is the factor for a pipe with ID = d1, Length = L1, and Friction Factor = f1. Friction factor f1 is computed using the curve fits for viscosity and density and Shacham's equation for friction factor. Note that K1 is a purely arbitrary symbol. If you want to us X1, Y1, alpha1, go ahead.

Now then, for pipe 2, dp = K2 x Q2^2 . The same pattern is used for all pipes connected.

Let us write the equations for a simple network.

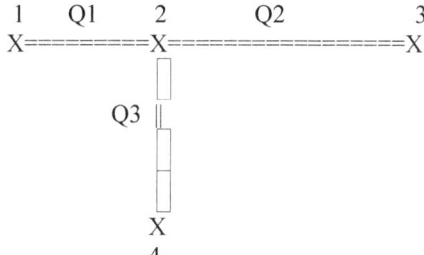

Let the flow be from 1 towards 2. It then splits with some flow towards 3 and some towards point 4. We wish to find out the flow rates for each pipe and pressures and each junction. From now on, the junction points will be called NODES. A node is defined as a point where the flow rate or pipe diameter changes. Pipes are just that. A length of tubing where the flow rate and pipe ID is constant.

For flow from 1 to 2, dp 1_2 = K1 x Q1^2
For flow from 2 to 3, dp 2_3 = K2 x Q2^2
For flow from 2 to 4, dp 2_4 = K3 x Q3^2

For this first problem, we will use Hazen-Williams (We walk before we run).
Now, for pressure drop in psi, remember that **hf = 4.52 x L x Q^1.85/(C^1.85 x d^4.87) and K = 4.52 x L /(C^1.85 x d^4.87)**

Pipe 1_2 has length L1, ft. and d1 in inches
Pipe 2_3 has length L2, ft. and d2 in inches
Pipe 2_4 has length L3, ft. and d3 in inches

Then, $K_1 = 4.52 \times L_1/(C^{1.85} \times d_1^{4.87})$
and $K_2 = 4.52 \times L_2/(C^{1.85} \times d_2^{4.87})$
and $K_3 = 4.52 \times L_3/(C^{1.85} \times d_3^{4.87})$

and $dp1\_2 = K_1 \times Q_1^{1.85}$, $dp2\_3 = K_2 \times Q_2^{1.85}$ and $dp2\_4 = K_3 \times Q_3^{1.85}$.

Also flow rate into point 2 has to equal flow into point 3 + flow into point 4. Then $Q_1 = Q_2 + Q_3$

Now, writing dp 1_2 and dp 2_3 and dp 2_4 is rather clumsy don't you think?
Realizing that dp 1_2 is the difference between pressure at 1 and at 2, we instead write:

$P_1 - P_2 = K_1 \times Q_1^{1.85}$, and
$P_2 - P_3 = K_2 \times Q_2^{1.85}$, and
$P_2 - P_4 = K_3 \times Q_3^{1.85}$, and
$Q_1 = Q_2 + Q_3$

Now, we have 4 equations and 7 unknowns (Q1, Q2, Q3, P1, P2, P3, P4)

This is the basic procedure for solving piping network equations.

Every node pressure and pipe flow rate is then calculated. It amounts to a system of non-linear equations.

There are 7 unknowns, and 4 equations. We must have 3 given quantities. Now, given P1 and Q1 and final P2 or P4, we then have 4 unknowns and 4 equations. This is solvable. Of course it can be done given Q3, P2 and P1 alternatively. The important thing is that the number of unknowns cannot exceed the number of equations, or no unique solution can be determined.

As we go along, more equations to help solve the networks will be developed. Now, let us solve the first problem. You will perhaps notice that for the pipe carrying flow Q1, the pipe constant is also called K1. You could call it K56 or K101 or whatever. BUT, if you do that, bookkeeping the equations becomes much more confusing. It is much easier to just use K1 for flow 1, K2 for flow 2, and so on. The pressure nodes can be numbered arbitrarily, but it is suggested that you start with the first input flow as P1. Then the first flow from P1 is naturally Q1. It is vital that you keep your data organized, in order to trouble shoot problems. Yes, if you are human, there will mistakes in inputting the data and you have to find the error(s). Don't make troubleshooting any harder than it already is.

First, we define the constants. These go into the RULES upper portion of TK SOLVER.

As a shortcut, let K=1.85. This reduces the number of keystrokes, an important thing when we get numerous equations to write. This will be input along with the numerical constants. Note also, that after writing the first one (for K1), it can be copied and then modified as seen below.

Also keep in mind that TK Solver is case-sensitive. That is K does NOT equal k, so keep your variables straight. The problems solved in this text were done with cap locks on all the time to avoid mixups. Engineering Suite 1.04 though, is NOT case-sensitive. In that program, k=K and c=C, and so on.

K1=4.52*(1/C)^K*L1*Q1^K*(D1^-4.8655) or, K1=4.52*Ll*(1/C)^K*(1/D1)^K if you like.
K2=4.52*(1/C)^K*L2*Q2^K*(D2^-4.8655)
K3=4.52*(1/C)^K*L3*Q3^K*(D3^-4.8655)

Now, we write the flow equations:

P1-P2=K1*Q1^K
P2-P3=K2*Q2^K
P2-P4=K3*Q3^K

Now the nodal equations (conservation of mass)

Q1+Q2=Q3

That is it for the RULES. Now go the the lower section for the variables or data.

Let L1 = 35', D1 = 1", L2 = 50', D2 = 1.25", L3= 67', D3 = 1.5".   These are just arbitrarily chosen, and for a real-world problem the actual pipe id would be input, depending on the type of pipe (steel, copper, cast iron, etc.)
Input C=130 (smooth piping), K=1.85 and all the pipe lengths and diameters.

Assume P1=20 psi, P3=2 psi, P2=5 psi. We assume a higher pressure drop in the smaller pipe. Input a guess of 1 for the unknowns. After all inputs are done, **SAVE the problem immediately before solving it.** The reason is that if the problem is not solved because of some error, you will have to input all the guesses again. For this first problem with just 4 unknowns, that is not too bad. Later, we will get to problems with over 40 unknowns. That would be a pain in the neck to re-input all those guesses again for another run. Plus, even if it solved perfectly the first time and you wanted to change a variable, such as a pipe size or length, if you did not have a saved version with all initial guesses, you would have to input them all over again. With a saved version, individual pipe lengths, diameters, flows and other variable can be easily changed and the problem re-run. Give it a descriptive name. For this first one, call it 1IN2OUT. When you get a solution you want to keep, call it lIN2OUT sol (for solution).

To print the rules, click on any line in the rules section, and go to FILE – PRINT and call it (Problem name) RULES. It is recommended to print it to a PDF. Note that the first page will be mostly blank. Delete it.

To print the solution, click on any line in the variables section, then FILE – PRINT and call it (Problem name) SOL. Note that if you input the units and a description of what each item is (gpm, pressure, final pressure, etc), print the file in LANDSCAPE mode so as to have a neat output.

The solution to this problem is Q_1=35.773548 gpm, Q_2=24.4270519 gpm, Q_3=11.3465598 gpm and P_2= 5.462 psi. As expected, Q_2 is greater than Q_3, being a larger pipe. As a check, compute P_1-P_2. By definition, $P_1-P_2=K_1*Q_1^K$ = 0.01942*35.7735^1.85 = 14.538 psi.
From results, P_2=5.4624 and Q_1=35.7735 gpm. Then given P_1=20, P_1-P_2 = 20-14.538 = 5.462 psi. Accurate to 0.005 psi. The comparison tolerances was set at 0.01. This is well with acceptable accuracy, given the tolerances of commercial pipe and using the HW equation with its built-in assumptions.

A point of interest is that the end pressures have to be reasonable. That is you cannot set P_3=19.99 psi and P_4=1 psi and expect results. That will produce an error. This a fact for the author has tried it.

Using the saved version of the RULES (you did save it, right?) one can easily change pipe sizes, lengths, and given quantities. You can input Q_1 instead of P_1 for example. You just **must have the same number of independent equations as unknowns**.

One could use another constant, say Z for 4.8655 (exponent for (1/D)) so that say P_1-P_2 = 4.52*L1+(1/C)^K*(1/D)^Z. This is merely a choice for the user to save keystrokes. But it does not matter since each line in the RULES section can be easily copied to the next line.

Another time saver the author has used is to set up templates for K1, K2, K3...etc and save as a template file. Then, just call it up, and then add the flow and nodal equations.

Rule

; 1 IN, 2 OUT

K1=4.52*(1/C)^K*L1*(D1^-4.8655)

K2=4.52*(1/C)^K*L2*(D2^-4.8655)

K3=4.52*(1/C)^K*L3*(D3^-4.8655)

P1-P2=K1*Q1^K

P2-P3=K2*Q2^K

P2-P4=K3*Q3^K

Q1=Q2+Q3

Status	Input	Name	Output	Unit	Comment
		K1	.019427248		C FACTOR FOR HW EQUATION
	130	C			EXPONENT USED IN HW EQUATION
	1.85	K			
	35	L1		FEET	
Guess	1	Q1		GPM	
	1	D1		INCHES	
		K2	.009371251		
	50	L2		FEET	
Guess	1	Q2		GPM	
	1.25	D2		INCHES	
		K3	.005171851		K FACTOR CALCULATED FOR PIPE 3, TYPICAL
	67	L3		FEET	
Guess	1	Q3		GPM	
	1.5	D3		INCHES	
	20	P1		PSI	
Guess	1	P2		PSI	
	2	P3		PSI	LOWER PRESSURE ASSUMED FOR SMALLER PIPE
	5	P4		PSI	HIGHER PRESSURE ASSUMED FOR LARGER PIPE

Variables

1IN2OUT.tkw

Status	Input	Name	Output	Unit	Comment
		K1	.019427248		C FACTOR FOR HW EQUATION
	130	C			C FACTOR FOR HW EQUATION
	1.85	K			EXPONENT USED IN HW EQUATION
	35	L1		FEET	
		Q1	35.773548	GPM	TOTAL FLOW
	1	D1		INCHES	
		K2	.009371251		
	50	L2		FEET	
		Q2	24.4279519	GPM	FLOW IN PIPE 2
	1.25	D2		INCHES	
		K3	.005171851		K FACTOR CALCULATED FOR PIPE 3, TYPICAL
	67	L3		FEET	
		Q3	11.345596	GPM	FLOW IN PIPE 3
	1.5	D3		INCHES	
	20	P1		PSI	
		P2	5.46244378	PSI	
	2	P3		PSI	LOWER PRESSURE ASSUMED FOR SMALLER PIPE
	5	P4		PSI	HIGHER PRESSURE ASSUMED FOR LARGER PIPE

Variables

05 SIMPLE NETWORKS

In this section, we will solve a simple network like the one that got this whole idea started. While working on a plumbing renovation for a large airport, the question was, what pressure to expect within the terminal, given the pressure upstream? See Diagram 5.01.

The problem is not as simple as it seems. The airport is served by a service line that runs alongside on the land side. It serves several valve rooms which reduce the pressure to 75 psig and then distribute it to an inner line running parallel to the long length of the terminal. So there are several inputs to a single line which in turn have several output lines. The minimum output at the end of the secondary lines is 25 psig to operate flush valves. We end up with 19 variables and 14 equations. The number of unknowns cannot exceed the number of equations. Thus, 5 variables must be given.
P1=75, P7=25, P8=25, Q7=60, and Q8=60. Note that any 5 could be given.
See sketch 5.01 and resulting solution.

Note that flow at P7 is about half that of Q7 and Q8, while flow at P10 (Q9) is excessive. That is to be expected, as P10 is much closer to the water source pressure.

What if there are 3 inputs and 2 outlets? See Diagram 5.02. This problem is done using the HW equation for feet of head. Heads for nodes P1, P5 and P9 are given and outlet node heads for P4 and P8 are given. For input, see 5.02 RULES and INPUT. For results, see 5.02 sol.

If you will notice, problem 5.02 is a cleverly disguised example of the infamous 3 Reservoir problem. Reservoir 1 is at node P1, reservoir 2 is at node P5 and reservoir 3 is at node P9. Isn't this an easy way to solve it?

You can play with these problems, changing input heads, pipe sizes, and lengths to see how the results change. Just be sure to save each variant to a different name or you will lose track of what you are doing.

No units were input in these problems. Feel free to do so, along with any comments. This can be easily done. It is understood that L is in feet, D is in inches, P is in feet of head or psi, and Q is in gpm.

You may notice that when a pipe turns 90 Deg., it is assigned a pressure point and the flow rate name is changed. There are 2 reasons for this.
1. Ease of bookkeeping. It makes the flow of writing the equations simpler. For a pipe with flow Qn, keep the K factor as Kn. For example, for Q1, the k factor = K1. For Q2, the k factor = K2.
2. If later, if you want to change the pipe where the flow turns 90 Deg. Into a header with a different size, this is easily done.

AIRPORT WATER SUPPLY (AWS)

5.1

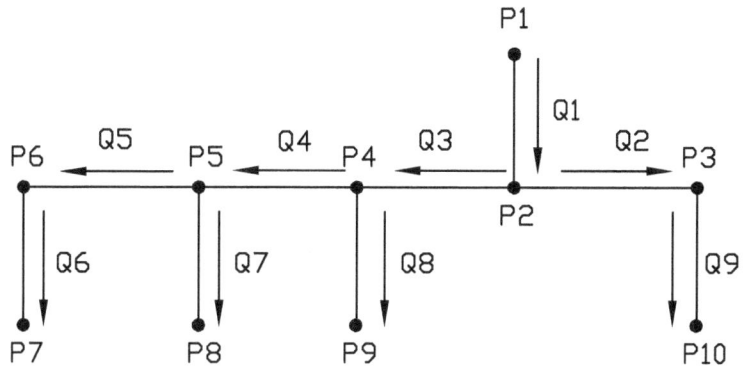

FLOW EQUATIONS:
$P1 - P2 = K1*Q1\wedge K$
$P2 - P3 = K2*Q2\wedge K$
$P2 - P4 = K3*Q3\wedge K$
$P4 - P5 = K4*Q4\wedge K$
$P5 - P6 = K5*Q5\wedge K$
$P6 - P7 = K6*Q6\wedge K$
$P5 - P8 = K7*Q7\wedge K$
$P4 - P9 = K8*Q8\wedge K$
$P3 - P10 = K9*Q9\wedge K$

NODES:
Q1 = Q2 + Q3
Q2 = Q9
Q3 = Q4 + Q8
Q4 = Q5 + Q7
Q5 = Q6

SET
P1 = 75 PSI
P7 = 25 PSI
P8 = 25 PSI
Q8 = 60 GPM
Q7 = 60 GPM

UNKNOWNS: (18)
P1, P2, P3, P4, P5, P6, P7, P8, P9, P10
Q1, Q2, Q3, Q4, Q5, Q6, Q7, Q8

Rule

$K1 = 4.52 * L1 * (1/C)^K * (1/D1)^{4.8655}$

$K2 = 4.52 * L2 * (1/C)^K * (1/D2)^{4.8655}$

$K3 = 4.52 * L3 * (1/C)^K * (1/D3)^{4.8655}$

$K4 = 4.52 * L4 * (1/C)^K * (1/D4)^{4.8655}$

$K5 = 4.52 * L5 * (1/C)^K * (1/D5)^{4.8655}$

$K6 = 4.52 * L6 * (1/C)^K * (1/D6)^{4.8655}$

$K7 = 4.52 * L7 * (1/C)^K * (1/D7)^{4.8655}$

$K8 = 4.52 * L8 * (1/C)^K * (1/D8)^{4.8655}$

$K9 = 4.52 * L9 * (1/C)^K * (1/D9)^{4.8655}$

$P1 - P2 = K1 * Q1^K$

$P2 - P3 = K2 * Q2^K$

$P2 - P4 = K3 * Q3^K$

$P4 - P5 = K4 * Q4^K$

$P5 - P6 = K5 * Q5^K$

$P6 - P7 = K6 * Q6^K$

$P5 - P8 = K7 * Q7^K$

$P4 - P9 = K8 * Q8^K$

$P3 - P10 = K9 * Q9^K$

$Q1 = Q2 + Q3$

$Q2 = Q9$

$Q3 = Q4 + Q8$

$Q4 = Q5 + Q7$

$Q5 = Q6$

Status	Input	Name	Output	Unit
		K1		
	250	L1		
	130	C		
	1.85	K		
	4	D1		
		K2		
	75	L2		
	2	D2		
		K3		
	100	L3		
	2	D3		
		K4		
	300	L4		
	2	D4		
		K5		
	300	L5		
	2	D5		
		K6		
	100	L6		
	2	D6		
		K7		
	100	L7		
	2	D7		
		K8		
	100	L8		
	2	D8		
		K9		
	100	L9		
	2	D9		
	75	P1		
Guess	1	P2		
Guess	1	Q1		
Guess	1	P3		
Guess	1	Q2		
Guess	1	P4		

Variables   5.01 AWS.tkw

Status	Input	Name	Output	Unit
Guess	1	Q3		
Guess	1	P5		
Guess	1	Q4		
Guess	1	P6		
Guess	1	Q5		
	25	P7		
Guess	1	Q6		
	25	P8		
	60	Q7		
Guess	1	P9		
	60	Q8		
Guess	1	P10		
Guess	1	Q9		

Status	Input	Name	Output	Unit
		K1	.0001632903	
	250	L1		
	130	C		
	1.85	K		
	4	D1		
		K2	.0014280488	
	75	L2		
	2	D2		
		K3	.0019040651	
	100	L3		
	2	D3		
		K4	.0057121953	
	300	L4		
	2	D4		
		K5	.0057121953	
	300	L5		
	2	D5		
		K6	.0019040651	
	100	L6		
	2	D6		
		K7	.0019040651	
	100	L7		
	2	D7		
		K8	.0019040651	
	100	L8		
	2	D8		
		K9	.0019040651	
	100	L9		
	2	D9		
	75	P1		
		P2	71.2779106	
		Q1	226.7680472	
		P3	66.7143324	
		Q2	78.4075776	
		P4	51.4799499	
		Q3	148.3604696	
		P5	28.7090407	
		Q4	88.3604695	
		P6	25.9272601	
		Q5	28.3604695	
	25	P7		
		Q6	28.3604695	
	25	P8		
	60	Q7		
		P9	47.7709092	
	60	Q8		
		P10	60.6295633	
		Q9	78.4075776	

CHAPTER 6 – SOME LOOPS

FIG 6.1 2 LOOPS

FIG. 6.2 2 LOOPS, DIVIDED FLOW, 1 OUTLET

FIG. 6.3 2 LOOPS, DIVIDED FLOW, 2 OUTLETS, TRIAL FLOWS

FIG. 6.4 2 LOOPS, DIVIDED FLOW, 2 OUTLETS, CORRECTED FLOWS

FIG. 6.5 3 LOOPS, SPRAY RINSE @ 60 DEG.F

FIG. 6.6 3 LOOPS, SPRAY RINSE @ 140 DEG.F

FIG. 6.7 3 LOOPS, TRIAL FLOWS

FIG. 6.8 3 LOOPS, CORRECTED FLOWS

FIG. 6.9 4 LOOPS, 4 HYDRANTS, TRIAL FLOWS

FIG. 6.10 4 LOOPS, 4 HYDRANTS, CORRECTED FLOWS

PROBLEM 6.1 – 2 LOOPS

The first problem is a simple 2-loop problem, where we are sure of the direction. It was solved using ESUITE.

2 LOOPS

FIG. 6.1

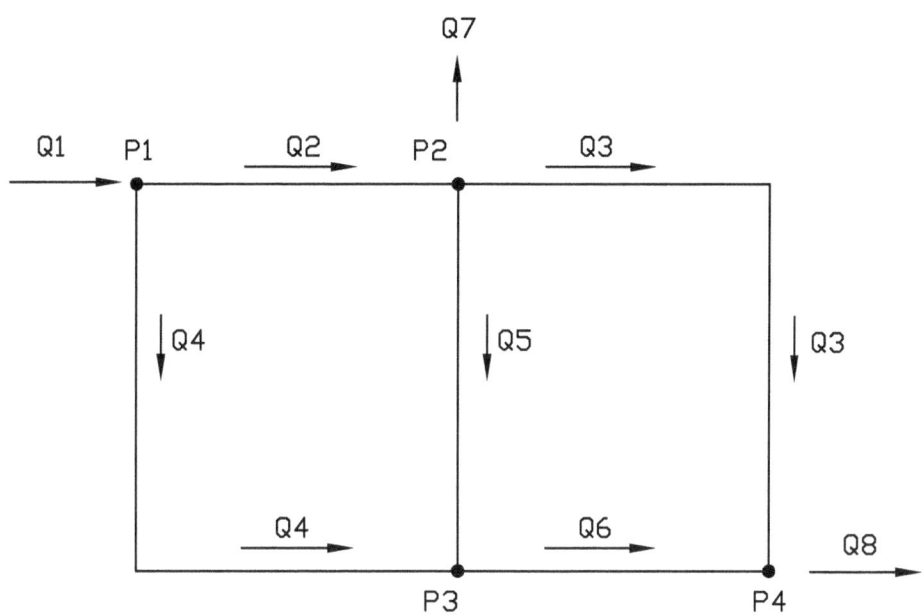

FLOW EQUATIONS:

$P1 - P2 = K1*Q2^K$
$P2 - P4 = K3*Q3^K$
$P1 - P3 = K4*Q4^K$
$P2 - P3 = K5*Q5^K$
$P3 - P4 = K6*Q6^K$

UNKNOWNS: (12)
P1, P2, P3, P4,
Q1, Q2, Q3, Q4, Q5, Q6, Q7, Q8

NODAL EQUATIONS:

$Q1 = Q2 + Q4$
$Q2 = Q7 + Q3 + Q5$
$Q4 + Q5 = Q6$
$Q6 = Q3 = Q8$

L2 = 300', D2 = 2"
L3 = 150', D3 = 2"
L4 = 300', D4 = 1"
L5 = 150', D5 = 2"
L6 = 400', D6 = 3"

Equations

```
/*   6.03 LOOPS   */
/*   ENGINEERING SUITE 1.04   */
K2=4.52*L2*C^-1.85*D2^-4.8655
K3=4.52*L3*C^-1.85*D3^-4.8655
K4=4.52*L4*C^-1.85*D4^-4.8655
K5=4.52*L5*C^-1.85*D5^-4.8655
K6=4.52*L6*C^-1.85*D6^-4.8655
P1-P2=K2*Q2^K
P2-P4=K3*Q3^K
P1-P3=K4*Q4^K
P2-P3=K5*Q5^K
P3-P4=K6*Q6^K
Q1=Q2+Q4
Q2=Q7+Q3+Q5
Q4+Q5=Q6
Q6+Q3=Q8
C=130
K=1.85
L2=300
D2=2
L3=150
D3=2
L4=300
D4=1
L5=150
D5=2
L6=400
D6=2
Q1=76
Q8=41
P1=100
```

Results

```
C=130
D2=2
```

```
D3=2
D4=1
D5=2
D6=2
K=1.85
K2=0.0057122
K3=0.0028561
K4=0.1665193
K5=0.0028561
K6=0.0076163
L2=300
L3=150
L4=300
L5=150
L6=400
P2=86.944697
P4=85.7597472
P3=86.9003078
P1=100
Q2=65.4134386
Q3=26.0073211
Q4=10.5865614
Q5=4.4061176
Q6=14.9926794
Q7=35
Q1=76
Q8=41
```

Log

```
"Variable"                              "Count"
C.......................................6
D2......................................2
D3......................................2
D4......................................2
D5......................................2
D6......................................2
K.......................................6
K2......................................2
K3......................................2
```

K4..2
K5..2
K6..2
L2..2
L3..2
L4..2
L5..2
L6..2
P2..3
P4..2
P3..3
P1..3
Q2..3
Q3..3
Q4..3
Q5..3
Q6..3
Q7..1
Q1..2
Q8..2

"Residuals"
C=130...............0
K=1.85..................0
L2=300..................0
D2=2................0
K2=4.52*L2*C^-1.85*D2^-4.8655............0
L3=150..................0
D3=2................0
K3=4.52*L3*C^-1.85*D3^-4.8655............0
L4=300..................0
D4=1................0
K4=4.52*L4*C^-1.85*D4^-4.8655............0
L5=150..................0
D5=2................0
K5=4.52*L5*C^-1.85*D5^-4.8655............0
L6=400..................0
D6=2................0
K6=4.52*L6*C^-1.85*D6^-4.8655............0
Q1=76................0
Q8=41................0

```
P1=100....................0
P1-P2=K2*Q2^K.................0
P2-P4=K3*Q3^K.................4.26326E-14
P1-P3=K4*Q4^K.................0
P2-P3=K5*Q5^K.................-7.10543E-14
P3-P4=K6*Q6^K.................-6.67253E-8
Q1=Q2+Q4..................0
Q2=Q7+Q3+Q5..................0
Q4+Q5=Q6..................-4.7411E-7
Q6+Q3=Q8..................4.7411E-7
```

PROBLEM 6.2 - DIVIDED FLOW – 2 LOOPS, DIVIDED FLOW, 1 INLET, 1 OUTLET

What happens when the flow is divided into 2 or more branches and then comes together again? We will see with this problem. See diagram 6.1. Note that the flow directions are obvious.

This was an early problem worked on, when the bugs were being worked out. Notice that the piping K factors do not match the flow rate. For instance, the pipe from P20 has flow rate Q4, but K factor of K5. The problem was still solved, but required more effort to keep the bookkeeping straight. That is the only instance of that in this text and is why it was not used again. You learn to avoid such errors.

Note how the flow divides, with the least flow in the smaller pipe, just as expected.

Diagram 6.1 is divided flow with 1 inlet, 1 outlet. What happens with 1 inlet, and 2 outlets? See diagram 6.2.

DIVIDED FLOW 1 OUTLET FIG. 6.2

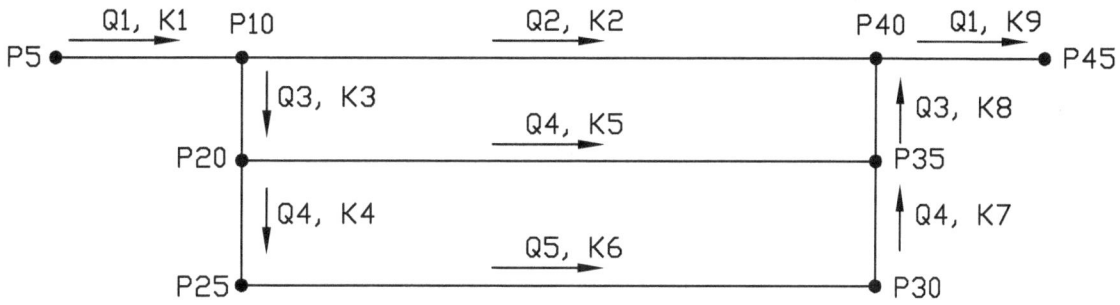

FLOW EQUATIONS:

P5 − P10 = K1*Q1^K
P10 − P20 = K2*Q2^K
P15 − P20 = K3*Q3^K
P20 − P25 = K4*Q5^K
P25 − P30 = K6*Q4^K
P30 − P35 = K7*Q5^K
P35 − P40 = K8*Q3^K
P40 − P45 = K9*Q1^K

NODAL EQUATIONS:
Q1 = Q2 + Q3
Q3 = Q4 = Q5

UNKNOWNS: (11)
P5, P10, P20, P25, P30, P35 P40, P45
Q1, Q2, Q3,

L1 = 90', D1 = 2"
L2 = 100', D2 = 1.25"
L3 = 50', D3 = 1.5"
L4 = 35', D4 = 1"
L5 = 100', D5 = 1"
L6 = 100', D6 = 1"
L7 = 10', D7 = 1.5"
L8 = 50', D8 = 1.5"
L9 = 40', D9= 2"

LET P5 = 60'
LET P45 = 2'

6-8

Rule

$K_1 = 0.002083 * L_1 * (100/C)^{1.85} * D_1^{-4.8655}$
$K_2 = 0.002083 * L_2 * (100/C)^{1.85} * D_2^{-4.8655}$
$K_3 = 0.002083 * L_3 * (100/C)^{1.85} * D_3^{-4.8655}$
$K_4 = 0.002083 * L_4 * (100/C)^{1.85} * D_4^{-4.8655}$
$K_5 = 0.002083 * L_5 * (100/C)^{1.85} * D_5^{-4.8655}$
$K_6 = 0.002083 * L_6 * (100/C)^{1.85} * D_6^{-4.8655}$
$K_7 = 0.002083 * L_7 * (100/C)^{1.85} * D_7^{-4.8655}$
$K_8 = 0.002083 * L_8 * (100/C)^{1.85} * D_8^{-4.8655}$
$K_9 = 0.002083 * L_9 * (100/C)^{1.85} * D_9^{-4.8655}$
$P_5 - P_{10} = K_1 * Q_1^K$
$P_{10} - P_{40} = K_2 * Q_2^K$
$P_{10} - P_{20} = K_3 * Q_3^K$
$P_{20} - P_{25} = K_4 * Q_5^K$
$P_{20} - P_{35} = K_5 * Q_4^K$
$P_{25} - P_{30} = K_6 * Q_5^K$
$P_{30} - P_{35} = K_7 * Q_5^K$
$P_{35} - P_{40} = K_8 * Q_3^K$
$P_{40} - P_{45} = K_9 * Q_1^K$
$Q_1 = Q_2 + Q_3$
$Q_3 = Q_4 + Q_5$

Status	Input	Name	Output	Unit
		K1	.003957994	
	90	L1		
	130	C		
	2	D1		
		K2	.043289084	
	100	L2		
	1.25	D2		
		K3	.008914398	
	50	L3		
	1.5	D3		
		K4	.044870624	
	35	L4		
	1	D4		
		K5	.128201783	
	100	L5		
	1	D5		
		K6	.128201783	
	100	L6		
	1	D6		
		K7	.001782879	
	10	L7		
	1.5	D7		
		K8	.008914398	
	50	L8		
	1.5	D8		
		K9	.001759108	
	40	L9		
	2	D9		
	60	P5		
Guess	1	P10		
Guess	1	Q1		
	1.85	K		
Guess	1	P40		
Guess	1	Q2		
Guess	1	P20		
Guess	1	Q3		
Guess	1	P25		
Guess	1	Q5		
Guess	1	P35		
Guess	1	Q4		
Guess	1	P30		
	2	P45		

Status	Input	Name	Output	Unit
		K1	.003957994	
	90	L1		
	130	C		
	2	D1		
		K2	.043289084	
	100	L2		
	1.25	D2		
		K3	.008914398	
	50	L3		
	1.5	D3		
		K4	.044870624	
	35	L4		
	1	D4		
		K5	.128201783	
	100	L5		
	1	D5		
		K6	.128201783	
	100	L6		
	1	D6		
		K7	.001782879	
	10	L7		
	1.5	D7		
		K8	.008914398	
	50	L8		
	1.5	D8		
		K9	.001759108	
	40	L9		
	2	D9		
	30	P5		
		P10	24.3616438	
		Q1	50.6629497	
	1.85	K		
		P40	4.505936075	
		Q2	27.4569578	
		P20	21.3662612	
		Q3	23.2059918	
		P25	17.8082983	
		Q5	10.6320213	
		P35	7.501318666	
		Q4	12.5739704	
		P30	7.642689999	
	2	P45		

PROBLEM 6.3 – 2 LOOPS, DIVIDED FLOW, 1 INLET, 2 OUTLETS, TRIAL

Notice that flow is assumed to go from P5 to P7 and from P7 to P8. It turns out that that assumption is incorrect. This is indicated by an exponentiation error at P5-P7. The flow (Q6) is actually flowing from P7 to P5 and is a negative. A negative number raised to 1.85 power produces an error. That is a tip off that the flow direction assumed is not right. The correction was made in 6.3 as well as correction to the nodal equations (cannot forget them). The problem then solved just fine. See 6.3 pages.

DIVIDED FLOW 2 OUTLETS FIG. 6.3

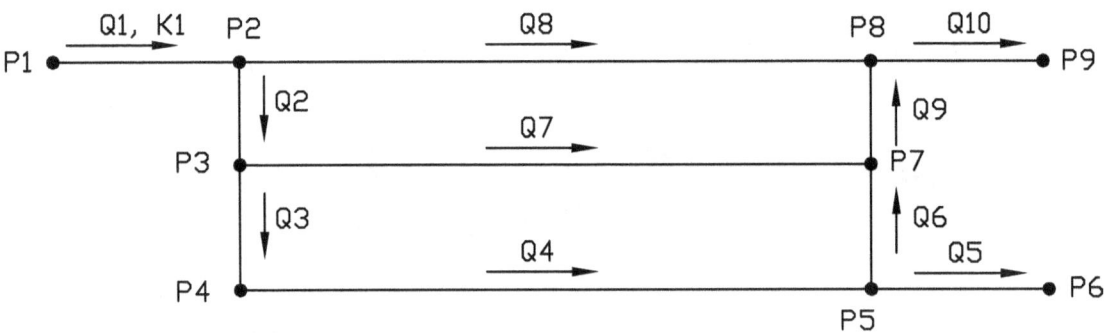

FLOW EQUATIONS:

P1 − P2 = K1*Q1^K
P2 − P3 = K2*Q2^K
P3 − P4 = K3*Q3^K
P4 − P5 = K4*Q4^K
P3 − P7 = K7*Q7^K
P2 − P8 = K8*Q8^K
P5 − P7 = K6*Q6^K
P7 − P8 = K9*Q9^K
P8 − P9 = K10*Q10^K
P5 − P6 = K5*Q5^K

NODAL EQUATIONS:

Q1 = Q2 + Q8
Q2 = Q7 + Q3
Q3 = Q4
Q4 = Q6 + Q5
Q7 = Q6 = Q9
Q8 + Q9 = Q10

UNKNOWNS: (19)
P1, P2, P3, P4, P5, P6, P7, P8, P9
Q1, Q2, Q3, Q4, Q5, Q6, Q7, Q8, Q9, Q10

L1 = 200', D1 = 4"
L2 = 300', D2 = 2"
L3 = 150', D3 = 2"
L4 = 300', D4 = 1"
L5 = 150', D5 = 2"
L6 = 400', D6 = 3"
L7 = 400', D7 = 2"
L8 = 400', D8 = 3"
L9 = 300', D9 = 2"
L10 = 200', D10 = 3"

LET P1 = 100 PSI
P9 = 50 PSI
P6 = 50 PSI

6-13

Rule

$K_1 = 4.52 * L_1 * C^{-1.85} * D_1^{-4.8655}$
$K_2 = 4.52 * L_2 * C^{-1.85} * D_2^{-4.8655}$
$K_3 = 4.52 * L_3 * C^{-1.85} * D_3^{-4.8655}$
$K_4 = 4.52 * L_4 * C^{-1.85} * D_4^{-4.8655}$
$K_5 = 4.52 * L_5 * C^{-1.85} * D_5^{-4.8655}$
$K_6 = 4.52 * L_6 * C^{-1.85} * D_6^{-4.8655}$
$K_7 = 4.52 * L_7 * C^{-1.85} * D_7^{-4.8655}$
$K_8 = 4.52 * L_8 * C^{-1.85} * D_8^{-4.8655}$
$K_9 = 4.52 * L_9 * C^{-1.85} * D_9^{-4.8655}$
$K_{10} = 4.52 * L_{10} * C^{-1.85} * D_{10}^{-4.8655}$

$P_1 - P_2 = K_1 * Q_1^K$
$P_2 - P_3 = K_2 * Q_2^K$
$P_3 - P_4 = K_3 * Q_3^K$
$P_4 - P_5 = K_4 * Q_4^K$
$P_3 - P_7 = K_7 * Q_7^K$
$P_2 - P_8 = K_8 * Q_8^K$
$P_5 - P_7 = K_6 * Q_6^K$
$P_7 - P_8 = K_9 * Q_9^K$
$P_8 - P_9 = K_{10} * Q_{10}^K$
$P_5 - P_6 = K_5 * Q_5^K$

$Q_1 = Q_2 + Q_8$
$Q_2 = Q_7 + Q_3$
$Q_3 = Q_4$
$Q_4 = Q_6 + Q_5$
$Q_7 + Q_6 = Q_9$
$Q_8 + Q_9 = Q_{10}$

Status	Input	Name	Output	Unit
	200	L1		
	130	C		
	4	D1		
	1.85	K		
		K1		
		K2		
	300	L2		
	2	D2		
		K3		
	150	L3		
	2	D3		
		K4		
	300	L4		
	1	D4		
		K5		
	150	L5		
	2	D5		
		K6		
	400	L6		
	3	D6		
		K7		
	400	L7		
	2	D7		
		K8		
	400	L8		
	3	D8		
		K9		
	300	L9		
	2	D9		
		K10		
	200	L10		
	3	D10		
	100	P1		
Guess	1	P2		
Guess	1	Q1		
Guess	1	P3		
Guess	1	Q2		
Guess	1	P4		
Guess	1	Q3		
Guess	1	P5		
		Q4		
Guess	1	P7		
Guess	1	Q7		
Guess	1	P8		
Guess	1	Q8		
Guess	1	Q6		
Guess	1	Q9		
	50	P9		
Guess	1	Q10		
	50	P6		

Variables   6.02.tkw

Status	Input	Name	Output	Unit
Guess	1	Q5		

Status	Input	Name	Output	Unit
	200	L1		
	130	C		
	4	D1		
	1.85	K		
		K1	.0001306322	
		K2	.0057121953	
	300	L2		
	2	D2		
		K3	.0028560976	
	150	L3		
	2	D3		
		K4	.166519266	
	300	L4		
	1	D4		
		K5	.0028560976	
	150	L5		
	2	D5		
> Error		K6	.00105918	
	400	L6		
	3	D6		
		K7	.0076162603	
	400	L7		
	2	D7		
		K8	.00105918	
	400	L8		
	3	D8		
		K9	.0057121953	
	300	L9		
	2	D9		
		K10	.00052959	
	200	L10		
	3	D10		
	100	P1		
Guess	96.5180729	P2		
Guess	187.5577178	Q1		
Guess	78.8504925	P3		
Guess	110.2219799	Q2		
Guess	78.4682664	P4		
Guess	14.57371735	Q3		
Guess	56.1833086	P5		
		Q4		
Guess	58.5514935	P7		
Guess	95.6482625	Q7		
Guess	61.5860135	P8		
Guess	269.84274	Q8		
Guess	-97.8347312	Q6		
Guess	-2.18646861	Q9		
	50	P9		
Guess	166.0834767	Q10		
	50	P6		

EXPONENT ERROR, FLOW IS IN OPPOSITE DIRECTION. (← pointing to K6 row)

PROBLEM 6.4 – 2 LOOPS, DIVIDED FLOW, 1 INLET, 2 OUTLETS, CORRECTED

Note that with the flow corrected, going from P7 to P5, the problem solved immediately.

This illustrates a limitation of this technique of solving network flows. **It will not automatically correct incorrect flow directions.** If the flow directions are correctly input, it will work. That is why problems with 3 or more loops are difficult to solve. Not impossible, just more effort.

DIVIDED FLOW 2 OUTLETS MODIFIED

FIG. 6.4

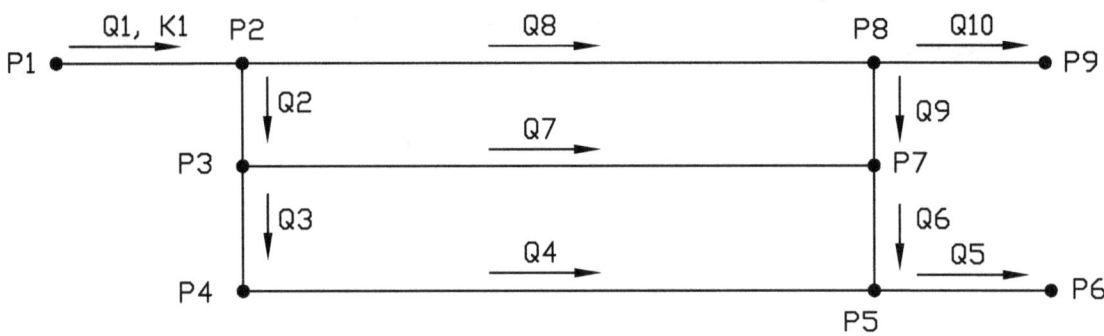

MODIFIED FLOW EQUATIONS:

$P1 - P2 = K1*Q1\wedge K$
$P2 - P3 = K2*Q2\wedge K$
$P3 - P4 = K3*Q3\wedge K$
$P4 - P5 = K4*Q4\wedge K$
$P3 - P7 = K7*Q7\wedge K$
$P2 - P8 = K8*Q8\wedge K$
$P7 - P5 = K6*Q6\wedge K$
$P8 - P7 = K9*Q9\wedge K$
$P8 - P9 = K10*Q10\wedge K$
$P5 - P6 = K5*Q5\wedge K$

MODIFIED NODAL EQUATIONS:
$Q1 = Q2 + Q8$
$Q2 = Q7 + Q3$
$Q3 = Q4$
$Q4 + Q6 = Q5$
$Q9 + Q7 = Q6$
$Q8 = Q9 + Q10$

UNKNOWNS: (19)
P1, P2, P3, P4, P5, P6, P7, P8, P9
Q1, Q2, Q3, Q4, Q5, Q6, Q7, Q8, Q9, Q10

L1 = 200', D1 = 4"
L2 = 300', D2 = 2"
L3 = 150', D3 = 2"
L4 = 300', D4 = 1"
L5 = 150', D5 = 2"
L6 = 400', D6 = 3"
L7 = 400', D7 = 2"
L8 = 400', D8 = 3"
L9 = 300', D9= 2"
L10 = 200', D10 = 3"

LET P1 = 100 PSI
P6 = 50 PSI
P8 = 50 PSI

Rule

$K1 = 4.52*L1*C\wedge-1.85*D1\wedge-4.8655$
$K2 = 4.52*L2*C\wedge-1.85*D2\wedge-4.8655$
$K3 = 4.52*L3*C\wedge-1.85*D3\wedge-4.8655$
$K4 = 4.52*L4*C\wedge-1.85*D4\wedge-4.8655$
$K5 = 4.52*L5*C\wedge-1.85*D5\wedge-4.8655$
$K6 = 4.52*L6*C\wedge-1.85*D6\wedge-4.8655$
$K7 = 4.52*L7*C\wedge-1.85*D7\wedge-4.8655$
$K8 = 4.52*L8*C\wedge-1.85*D8\wedge-4.8655$
$K9 = 4.52*L9*C\wedge-1.85*D9\wedge-4.8655$
$K10 = 4.52*L10*C\wedge-1.85*D10\wedge-4.8655$

P1-P2=K1*Q1^K
P2-P3=K2*Q2^K
P3-P4=K3*Q3^K
P4-P5=K4*Q4^K
P3-P7=K7*Q7^K
P2-P8=K8*Q8^K
P7-P5=K6*Q6^K
P8-P7=K9*Q9^K
P8-P9=K10*Q10^K
P5-P6=K5*Q5^K

Q1=Q2+Q8
Q2=Q7+Q3
Q3=Q4
Q4+Q6=Q5
Q9+Q7=Q6
Q8=Q9+Q10
Q1=Q10+Q5

Status	Input	Name	Output	Unit
	200	L1		
	130	C		
	4	D1		
	1.85	K		
		K1		
		K2		
	300	L2		
	2	D2		
		K3		
	150	L3		
	2	D3		
		K4		
	300	L4		
	1	D4		
		K5		
	150	L5		
	2	D5		
		K6		
	400	L6		
	3	D6		
		K7		
	400	L7		
	2	D7		
		K8		
	400	L8		
	3	D8		
		K9		
	300	L9		
	2	D9		
		K10		
	200	L10		
	3	D10		
	100	P1		
Guess	1	P2		
Guess	1	Q1		
Guess	1	P3		
Guess	1	Q2		
Guess	1	P4		
Guess	1	Q3		
Guess	1	P5		
Guess	1	Q4		
Guess	1	P7		
Guess	1	Q7		
Guess	1	P8		
Guess	1	Q8		
Guess	1	Q6		
Guess	1	Q9		
	50	P9		
Guess	1	Q10		
	50	P6		

Variables

Status	Input	Name	Output	Unit
Guess	1	Q5		

Status	Input	Name	Output	Unit
	200	L1		
	130	C		
	4	D1		
	1.85	K		
		K1	.0001306322	
		K2	.0057121953	
	300	L2		
	2	D2		
		K3	.0028560976	
	150	L3		
	2	D3		
		K4	.166519266	
	300	L4		
	1	D4		
		K5	.0028560976	
	150	L5		
	2	D5		
		K6	.00105918	
	400	L6		
	3	D6		
		K7	.0076162602	
	400	L7		
	2	D7		
		K8	.00105918	
	400	L8		
	3	D8		
		K9	.0057121953	
	300	L9		
	2	D9		
		K10	.00052959	
	200	L10		
	3	D10		
	100	P1		
		P2	94.0844373	
		Q1	328.6452904	
		P3	78.2528273	
		Q2	72.5990324	
		P4	77.9523812	
		Q3	12.3870937	
		P5	60.4354542	
		Q4	12.3870937	
		P7	63.3195678	
		Q7	60.2119387	
		P8	63.8599557	
		Q8	256.046258	
		Q6	71.9084260	
		Q9	11.6964873	
	50	P9		
		Q10	244.3497706	
	50	P6		

Status	Input	Name	Output	Unit
		Q5	84.2955197	

PROBLEM 6.5 3 LOOPS, SPRAY RINSE @ 60 Deg.F

This problem comes directly from the book, "2500 Fluid Mechanics Solved" by Schaums. It is supposed to represent a rinse wash system. See diagram 6.03.
In this problem, the actual friction factors will be calculated for each pipe. First, the viscosity and density of the water is calculated (based upon a given temperature of 60 Deg.F), using the curve fits previously discussed. Then the Reynolds number for each pipe is calculated (with an assumed flow). Then, with the Reynolds number and the roughness factor (based upon the piping material, the friction factor is computed. It is very remarkable how fast TK does this, especially since Re is based upon the flow rate which is unknown. An assumed flow rate is used (initial guess is 1 for each pipe) and from that the rest is computed. The resulting residuals from the flow and nodal equations is compared to the accuracy desired. If not accurate enough, the whole process starts again until it does. All in a fraction of a second.

3 LOOPS

FIG. 6.5

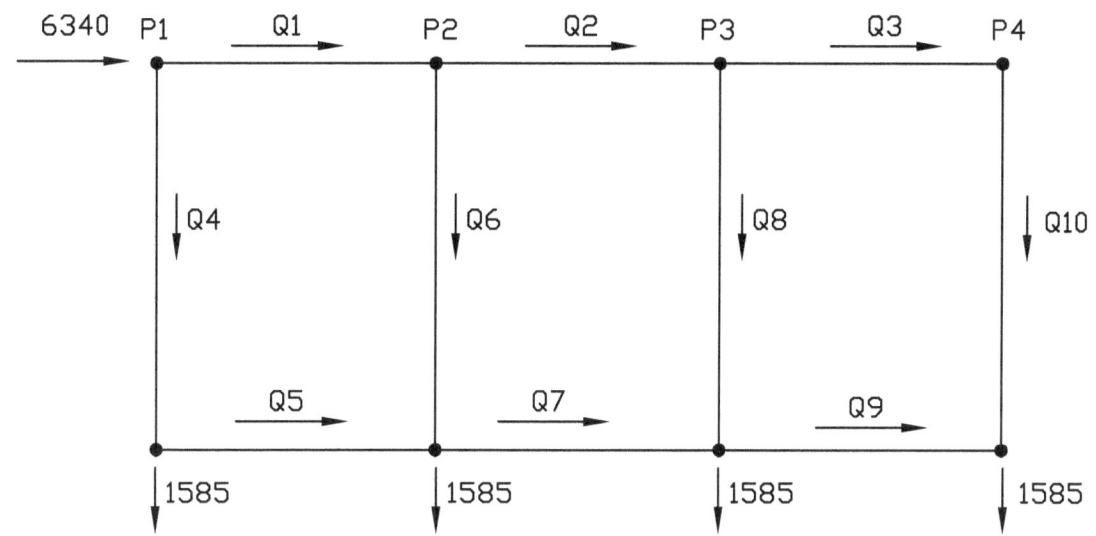

FLOW EQUATIONS:

P1 - P2 = C1*Q2^K
P2 - P3 = C2*Q2^K
P3 - P4 = C3*Q3^K
P1 - P5 = C4*Q4^K
P2 - P6 = C6*Q6^K
P3 - P7 = C8*Q8^K
P4 - P8 = C10*Q10^K
P5 - P6 = C5*Q5^K
P6 - P7 = C7*Q7^K
P7 - P8 = C9*Q9^K

UNKNOWNS: (18)
P1, P2, P3, P4, P5, P6, P7, P8
Q1, Q2, Q3, Q4, Q5, Q6, Q7, Q8, Q9 Q10

NODAL EQUATIONS:

6340 = Q1 + Q4
Q1 = Q6 + Q2
Q2 = Q8 + Q3
Q3 = Q10
Q4 = 1585 + Q5
Q5 + Q6 = 1585 + Q7
Q7 + Q8 = Q9 + 1585
Q9 = Q10 = 1585

L1 = 1968.5' D1 = 12"
L2 = 1968.5', D2 = 12"
L3 = 1968.5', D3 = 12"
L4 = 1312', D4 = 10"
L5 = 1968.5', D5 = 12"
L6 = 1312', D6 = 10"
L7 = 1968.5', D7 = 12"
L8 = 1312', D8 = 10"
L9 = 1968.5', D9 = 12"
L10 = 1312', D10 = 10"

GIVEN:
P1 = 100' W.G.

6-26

Rule
v=a1+a2*T+a3*T^2+a4*T^3+a5*T^4+a6*T^5

rho=b1+b2*T+b3*T^2+b4*T^3

R1=50.6*Q1*rho/(D1*v)
R2=50.6*Q3*rho/(D2*v)
R3=50.6*Q3*rho/(D3*v)
R4=50.6*Q4*rho/(D4*v)
R5=50.6*Q5*rho/(D5*v)
R6=50.6*Q6*rho/(D6*v)
R7=50.6*Q7*rho/(D7*v)
R8=50.6*Q8*rho/(D8*v)
R9=50.6*Q9*rho/(D9*v)
R10=50.6*Q10*rho/(D10*v)

Z1=3.243*e/D1
Z2=3.243*e/D2
Z3=3.243*e/D3
Z4=3.243*e/D4
Z5=3.243*e/D5
Z6=3.243*e/D6
Z7=3.243*e/D7
Z8=3.243*e/D8
Z9=3.243*e/D9
Z10=3.243*e/D10

f1=(-2*log(Z1-(5.02/R1)*log(Z1+14.5/R1)))^-2
f2=(-2*log(Z2-(5.02/R2)*log(Z2+14.5/R2)))^-2
f3=(-2*log(Z3-(5.02/R3)*log(Z3+14.5/R3)))^-2
f4=(-2*log(Z4-(5.02/R4)*log(Z4+14.5/R4)))^-2
f5=(-2*log(Z5-(5.02/R5)*log(Z5+14.5/R5)))^-2
f6=(-2*log(Z6-(5.02/R6)*log(Z6+14.5/R6)))^-2
f7=(-2*log(Z7-(5.02/R7)*log(Z7+14.5/R7)))^-2
f8=(-2*log(Z8-(5.02/R8)*log(Z8+14.5/R8)))^-2
f9=(-2*log(Z9-(5.02/R9)*log(Z9+14.5/R9)))^-2
f10=(-2*log(Z10-(5.02/R10)*log(Z10+14.5/R10)))^-2

VEL1=.4085*Q1/(D1^2)
VEL2=.4085*Q2/(D2^2)
VEL3=.4085*Q3/(D3^2)
VEL4=.4085*Q4/(D4^2)
VEL5=.4085*Q5/(D5^2)
VEL6=.4085*Q6/(D6^2)
VEL7=.4085*Q7/(D7^2)
VEL8=.4085*Q8/(D8^2)
VEL9=.4085*Q9/(D9^2)
VEL10=.4085*Q10/(D10^2)

C1=0.03108*f1*L1/(D1^5)
C2=0.03108*f2*L2/(D2^5)

Rule
$C3 = 0.03108 \cdot f3 \cdot L3 / (D3^5)$
$C4 = 0.03108 \cdot f4 \cdot L4 / (D4^5)$
$C5 = 0.03108 \cdot f5 \cdot L5 / (D5^5)$
$C6 = 0.03108 \cdot f6 \cdot L6 / (D6^5)$
$C7 = 0.03108 \cdot f7 \cdot L7 / (D7^5)$
$C8 = 0.03108 \cdot f8 \cdot L8 / (D8^5)$
$C9 = 0.03108 \cdot f9 \cdot L9 / (D9^5)$
$C10 = 0.03108 \cdot f10 \cdot L10 / (D10^5)$

$P1 - P2 = C1 \cdot Q1^2$
$P2 - P3 = C2 \cdot Q2^2$
$P3 - P4 = C3 \cdot Q3^2$
$P1 - P5 = C4 \cdot Q4^2$
$P2 - P6 = C5 \cdot Q5^2$
$P3 - P7 = C6 \cdot Q6^2$
$P4 - P8 = C7 \cdot Q7^2$
$P5 - P6 = C8 \cdot Q8^2$
$P6 - P7 = C9 \cdot Q9^2$
$P7 - P8 = C10 \cdot Q10^2$

$Q11 = Q1 + Q4$
$Q1 = Q5 + Q2$
$Q2 = Q6 + Q3$
$Q4 = Q8 + Q12$
$Q8 + Q5 = Q9 + Q13$
$Q9 + Q6 = Q10 + Q14$
$Q7 = Q3$
$Q7 + Q10 = Q15$

Status	Input	Name	Output	Unit	Comment
		v		CENTIPOIS	VISCOSITY OF WATER AT TEMP T
	.000557280	a1			CURVE FIT CONSTANT
	.22430157	a2			
	60	T		DEG.F	TEMPERATURE OF THE WATER, DEG.F
	-.0091107	a3			
	.000155423	a4			
	-1.23944E-6	a5			
	3.790416E-9	a6			
		rho		LB/CU.FT.	DENSITY OF WATER AT TEMP T
	62.2614555	b1			CURVE FIT CONSTANT
	.009339695	b2			
	-.000137041	b3			
	1.83106E-7	b4			
		R1			REYNOLDS NO. FOR PIPE 1
Guess	1	Q1		GPM	FLOW TO BE CALCULATED
	12	D1			PIPE 1 DIAMETER, IN.
		R2			
Guess	1	Q3			
	12	D2			
		R3			
	12	D3			
		R4			
Guess	1	Q4			
	10	D4			
		R5			
Guess	1	Q5			
	10	D5			
		R6			
Guess	1	Q6			
	10	D6			
		R7			
Guess	1	Q7			
	10	D7			
		R8			
Guess	1	Q8			
	12	D8			
		R9			

Status	Input	Name	Output	Unit	Comment
Guess	1	Q9			
	12	D9			
		R10			
Guess	1	Q10			
	12	D10			
		Z1			
	.00015	e		FEET	PIPE ROUGHNESS, FT.
		Z2			
		Z3			
		Z4			
		Z5			
		Z6			
		Z7			
		Z8			
		Z9			
		Z10			
		f1			CALCULATED FRICTION FACTOR 1
		f2			
		f3			
		f4			
		f5			
		f6			
		f7			
		f8			
		f9			
		f10			
		VEL1			WATER VELOCITY, FT/SEC IN PIPE 1
		VEL2			
Guess	1	Q2		GPM	FLOW TO BE CALCULATED
		VEL3			
		VEL4			
		VEL5			
		VEL6			
		VEL7			
		VEL8			
		VEL9			
		VEL10			

Variables — dw shacham SPRAY RINSE.tkw

Variables

Status	Input	Name	Output	Unit	Comment
		C1			
	1968.5	L1			PIPE 1 LENGTH, FT
		C2			
	1968.5	L2			
		C3			
	1968.5	L3			
		C4			
	1312.3	L4			
		C5			
	1312.3	L5			
		C6			
	1312.3	L6			
		C7			
	1312.3	L7			
		C8			
	1968.5	L8			
		C9			
	1968.5	L9			
		C10			
	1968.5	L10			
	100	P1		FT. W.G.	GIVEN INPUT PRESSURE, FT. OF WATER
Guess	1	P2		FT. W.G.	CALCULATED PRESSURE @ P2
Guess	1	P3			
Guess	1	P4			
Guess	1	P5			
Guess	1	P6			
Guess	1	P7			
Guess	1	P8			
Guess	1600	Q11		GPM	ASSUMED INPUT GPM
	1585	Q13		GPM	GIVEN OUTFLOW, GPM
	1585	Q14		GPM	GIVEN OUTFLOW, GPM
	1585	Q15		GPM	GIVEN OUTFLOW, GPM
	1585	Q12		GPM	GIVEN OUTFLOW, GPM

Status	Input	Name	Output	Unit	Comment
		v	1.11596342	CENTIPOIS	VISCOSITY OF WATER AT TEMP T
	.000557280	a1			CURVE FIT CONSTANT
	.22430157	a2			
	60	T		DEG.F	TEMPERATURE OF THE WATER, DEG.F
	-.0091107	a3			
	.000155423	a4			
	-1.23944E-6	a5			
	3.790416E-9	a6			
		rho	62.3680405	LB/CU.FT.	DENSITY OF WATER AT TEMP T
	62.2614555	b1			CURVE FIT CONSTANT
	.009339695	b2			
	-.000137041	b3			
	1.83106E-7	b4			
		R1	811425.663		REYNOLDS NO. FOR PIPE 1
		Q1	3443.24027	GPM	FLOW TO BE CALCULATED
	12	D1			PIPE 1 DIAMETER, IN.
		R2	211017.455		
		Q3	895.440991		
	12	D2			
		R3	211017.455		
		Q4	2896.75973		
	12	D3			
		R4	819172.059		
	10	D4			
		R5	437217.499		
		Q5	1546.09038		
	10	D5			
		R6	283272.350		
		Q6	1001.70889		
	10	D6			
		R7	253220.946		
		Q7	895.440991		
	10	D7			
		R8	309126.121		
		Q8	1311.75973		
	12	D8			
		R9	299956.775		

Status	Input	Name	Output	Unit	Comment
		Q9	1272.850114		
	12	D9			
		R10	162499.806		
		Q10	689.559008		
	12	D10			
	.00015	Z1	.0000405371	FEET	PIPE ROUGHNESS, FT.
		e			
		Z2	.0000405371		
		Z3	.0000405371		
		Z4	.000048645		
		Z5	.000048645		
		Z6	.000048645		
		Z7	.000048645		
		Z8	.0000405371		
		Z9	.0000405371		
		Z10	.0000405371		
		f1	.0143378571		CALCULATED FRICTION FACTOR 1
		f2	.0166891771		
		f3	.0166891771		
		f4	.0146617491		
		f5	.0154724931		
		f6	.0162487531		
		f7	.0164816331		
		f8	.0158389631		
		f9	.0159002931		
		f10	.0173657601		
		VEL1	9.767803121		WATER VELOCITY, FT/SEC IN PIPE 1
		VEL2	5.381845331		
		Q2	1897.149881	GPM	FLOW TO BE CALCULATED
		VEL3	2.540191971		
		VEL4	11.8332635		
		VEL5	6.315779211		
		VEL6	4.091980831		
		VEL7	3.65787645		
		VEL8	3.721207291		
		VEL9	3.610828271		
		VEL10	1.956144821		

Variables

Status	Input	Name	Output	Unit	Comment
		C1	3.525287E-0		
	1968.5	L1			PIPE 1 LENGTH, FT
		C2	4.103412E-0		
	1968.5	L2			
		C3	4.103412E-0		
	1968.5	L3			
		C4	5.979983E-0		
	1312.3	L4			
		C5	6.310655E-0		
	1312.3	L5			
		C6	6.627263E-0		
	1312.3	L6			
		C7	6.722246E-0		
	1312.3	L7			
		C8	3.894368E-0		
	1968.5	L8			
		C9	3.909447E-0		
	1968.5	L9			
		C10	4.269765E-0		
	1968.5	L10			
	100	P1		FT. W.G.	GIVEN INPUT PRESSURE, FT. OF WATER
		P2	58.2045382	FT. W.G.	CALCULATED PRESSURE @ P2
		P3	43.4356289		
		P4	40.1454533		
		P5	49.8206683		
		P6	43.1195767		
		P7	36.7856962		
		P8	34.7554585		
		Q11	6340	GPM	ASSUMED INPUT GPM
	1585	Q13		GPM	GIVEN OUTFLOW, GPM
	1585	Q14		GPM	GIVEN OUTFLOW, GPM
	1585	Q15		GPM	GIVEN OUTFLOW, GPM
	1585	Q12		GPM	GIVEN OUTFLOW, GPM

Variables

dw shacham SPRAY RINSE.tkw

PROBLEM 6.6 3 LOOPS, SPRAY RINSE @ 140 Deg.F

You might be thinking, yeah, but what if the water is not at 61 Deg. F like the HW equation assumes? In that case, we will use the Shacham equation for friction factor and Darcy-Weisbach for pressure drop.

Darcy-Weisbach: **hf= f x (L/D) x V^2/(2 x g)**

Shacham: **f = {-2 x log(e/D/(3.7) – (5.02/Re) x log(e/D/(3.7) + 14.5/Re))}**

Now this takes into account different temperature of the fluid. In most of these problems, it will be water. Fortunately, we already have curve fits to determine the viscosity and density of water at temperature.

Note the Z factor. This merely an attempt to simplify the Shacham equation since the value of 3.243*e/D occurs twice in the equation. It really does not save keystrokes, so won't be used again.

Compare the differences for HW to the Darcy equation results. There is a considerable difference. Note the Reynolds numbers. They all show turbulent flow as predicted earlier. Obviously, the Darcy-Weisbach equation using Shacham's equation for friction factor is more accurate. It will be used mostly, except for Fire Sprinkler problems. For that, HW will be used, as that is what NFPA 13

Status	Input	Name	Output	Unit	Comment
		v	7.03054737	CENTIPOISE	VISCOSITY OF WATER AT TEMP T
	.000557280	a1			CURVE FIT CONSTANT
	.22430157	a2			
	140	T		DEG.F	TEMPERATURE OF THE WATER, DEG.F
	-.0091107	a3			
	.000155423	a4			
	-1.23944E-6	a5			
	3.790416E-9	a6			
		rho	61.3854520	LB/CU.FT.	DENSITY OF WATER AT TEMP T
	62.2614555	b1			CURVE FIT CONSTANT
	.009339695	b2			
	-.000137041	b3			
	1.83106E-7	b4			
		R1	126181.201		REYNOLDS NO. FOR PIPE 1
		Q1	3427.27591	GPM	FLOW TO BE CALCULATED
	12	D1			PIPE 1 DIAMETER, IN.
		R2	32588.0926		
		Q3	885.142822		
	12	D2			
		R3	32588.0926		
		D3			
	12				
		R4	128684.483		
		Q4	2912.72408		
	10	D4			
		R5	68754.2919		
		Q5	1556.22711		
	10	D5			
		R6	43557.4390		
		Q6	985.905982		
	10	D6			
		R7	39105.7111		
		Q7	885.142822		
	10	D7			
		R8	48882.5014		
		Q8	1327.72408		
	12	D8			
		R9	47823.1769		

Status	Input	Name	Output	Unit	Comment
		Q9	1298.951195		
	12	D9			
		R10	25766.47513		
		Q10	699.857177		
	12	D10			
	.00015	Z1	.000040537	FEET	PIPE ROUGHNESS, FT.
		e			
		Z2	.000040537		
		Z3	.000040537		
		Z4	.000048645		
		Z5	.000048645		
		Z6	.000048645		
		Z7	.000048645		
		Z8	.000040537		
		Z9	.000040537		
		Z10	.000040537		
		f1	.018098908		CALCULATED FRICTION FACTOR 1
		f2	.023540422		
		f3	.023540422		
		f4	.018197967		
		f5	.020315201		
		f6	.022224024		
		f7	.022725164		
		f8	.021612465		
		f9	.021709491		
		f10	.024794304		
		VEL1	9.722251536		WATER VELOCITY, FT/SEC IN PIPE 1
		VEL2	5.30780164		
		Q2	1871.048805	GPM	FLOW TO BE CALCULATED
		VEL3	2.51097807		
		VEL4	11.8984778		
		VEL5	6.35718776		
		VEL6	4.02742593		
		VEL7	3.61580843		
		VEL8	3.766495049		
		VEL9	3.68487196		
		VEL10	1.985358725		

Variables

dw shacham SPRAY RINSE.tkw

Status	Input	Name	Output	Unit	Comment
		C1	4.450026E-6		
	1968.5	L1			PIPE 1 LENGTH, FT
		C2	5.787946E-6		
	1968.5	L2			
		C3	5.787946E-6		
	1968.5	L3			
		C4	7.422275E-6		
	1312.3	L4			
		C5	8.285816E-6		
	1312.3	L5			
		C6	9.064354E-6		
	1312.3	L6			
		C7	9.26875E-6		
	1312.3	L7			
		C8	5.313914E-6		
	1968.5	L8			
		C9	5.33777E-6		
	1968.5	L9			
		C10	6.096241E-6		
	1968.5	L10			
	100	P1		FT. W.G.	GIVEN INPUT PRESSURE, FT. OF WATER
		P2	47.7290097	FT. W.G.	CALCULATED PRESSURE @ P2
		P3	27.4664325		
		P4	22.9317055		
		P5	37.0297057		
		P6	27.6620659		
		P7	18.6557845		
		P8	15.6698452		
		Q11	6340	GPM	ASSUMED INPUT GPM
	1585	Q13		GPM	GIVEN OUTFLOW, GPM
	1585	Q14		GPM	GIVEN OUTFLOW, GPM
	1585	Q15		GPM	GIVEN OUTFLOW, GPM
	1585	Q12		GPM	GIVEN OUTFLOW, GPM

Variables

dw shacham SPRAY RINSE.tkw

PROBLEM 6.7 – 3 LOOPS, TRIAL FLOWS

This is a loop problem with 3 loops and we cannot be certain of the directions of the flows. We must guess. Thankfully, there is a neat way to find the corrections required. We write loop equations, 1 for each loop, assuming flow rates. The assumed flow rates are a guesstimate, but still must obey conservation of mass. That is, what goes into a node must come out and there is no net gain or loss of fluid in the system. Our assumed flow rates are probably not correct, so will need a correction, 1/loop. For instance, for loop 1, the correction = DQ1 (meaning delta Q, loop 1). The loop equations are familiar if you have used the Hardy Cross Method. What it amounts to is that the total pressure drop in a loop = 0. Flows clockwise are considered positive (+) and flows counterclockwise are considered negative (-). For loop 1, DQ1 is added for clockwise flow, and subtracted for counterclockwise flow. For common loops, say loop 1 and loop 2, if the flow is (+) for loop 1 and (-) for loop 2, then the correction factor = DQ1 – DQ2. See diagram 6.05.

For loop 1, the loop equation is: $K1*Q1^K + K7*Q7^K + K9*Q9^K – K8*Q8^K – K2*Q2^K = 0$

For loop 2, the loop equation is: $K8*Q8^K + K10*Q10^K – K4*Q4^K – K3*Q3^K = 0$

For loop 3, the loop equation is: $K6*Q6^K – K5*Q5^K – K10*Q10^K – K9*Q9^K = 0$

Okay so far? Q1 is in loop 1, Q2 is in loop 1, Q3 is in loop 2, Q4 is in loop 2, Q5 is in loop 3, Q6 is in loop 3, Q7 is in loop 1, Q8 is in loops 1 AND 2, Q9 is in loops 1 AND 3 and Q10 is in loops 2 AND 3.

Then, corrected flows are Q1 = 500 + DQ1, Q2 = 500 – DQ1 (Q2 is going counterclockwise), Q3 = 300 – DQ3, Q4 = 100 – DQ4, Q5 = 250 – DQ3, Q6 = 100 + DQ3, Q7 = 500 + DQ1, Q8 = 200 – DQ1 + DQ2, Q9 = 400 + DQ1 – DQ3, and Q10 = 150 – DQ3 + DQ2.

For loop 1, the loop equation is: $K1*(Q1 + DQ1)^K + K7*(Q7 + DQ1)^K + K9*(400 + DQ2 – DQ3)^K – K8*(200 – DQ1 + DQ2)^K – K2*(500 – DQ1)^K = 0$

For loop 2, the loop equation is: $K8*(200 + DQ2 – DQ1)^K + K10*(150 + DQ1 – DQ3)^K – K4*(100 – DQ2)^K – K3*(300 – DQ2)^K = 0$

For loop 3, the loop equation is: $K6*(100 + DQ3)^K – K5*(250 – DQ3)^K – K10*(150 – DQ3 + DQ2)^K – K9*(400 – DQ3 + DQ2)^K = 0.$

We are then presented with 3 equations, 3 unknowns (DQ1, DQ2, and DQ3). TK or Esuite can easily handle this.

Plug in the corrections to determine the flow rates. Even better, let TK do that for you. For example after computing DQ1, then find the corrected Q1 as Q! = 500 + DQ1 and so on. What we are really after is the directions of the flows. If all the flows are positive, then we guessed correctly. Then we write the flow equations as normal, that is, P1 – P2 = K1*Q1^K and so on to compute the pressures and flows. Now it could turn out that some of the flows show up as negative. This means that those flows are actually flowing in the opposite directions assumed. In that case, the flow diagram must be revised to show the corrected flow directions. The flow and nodal equations must also reflect this correction. After input and solution (set accuracy = 0.1), the results are shown. Note that Q4 is Negative. That means that the flow is actually from P4 to P3. The corrected diagram is 6.8 3 LOOPS CORRECTED FLOWS. Note the reverse flow for Q4 and resulting flow and nodal equations.

3 LOOPS
TRIAL FLOWS

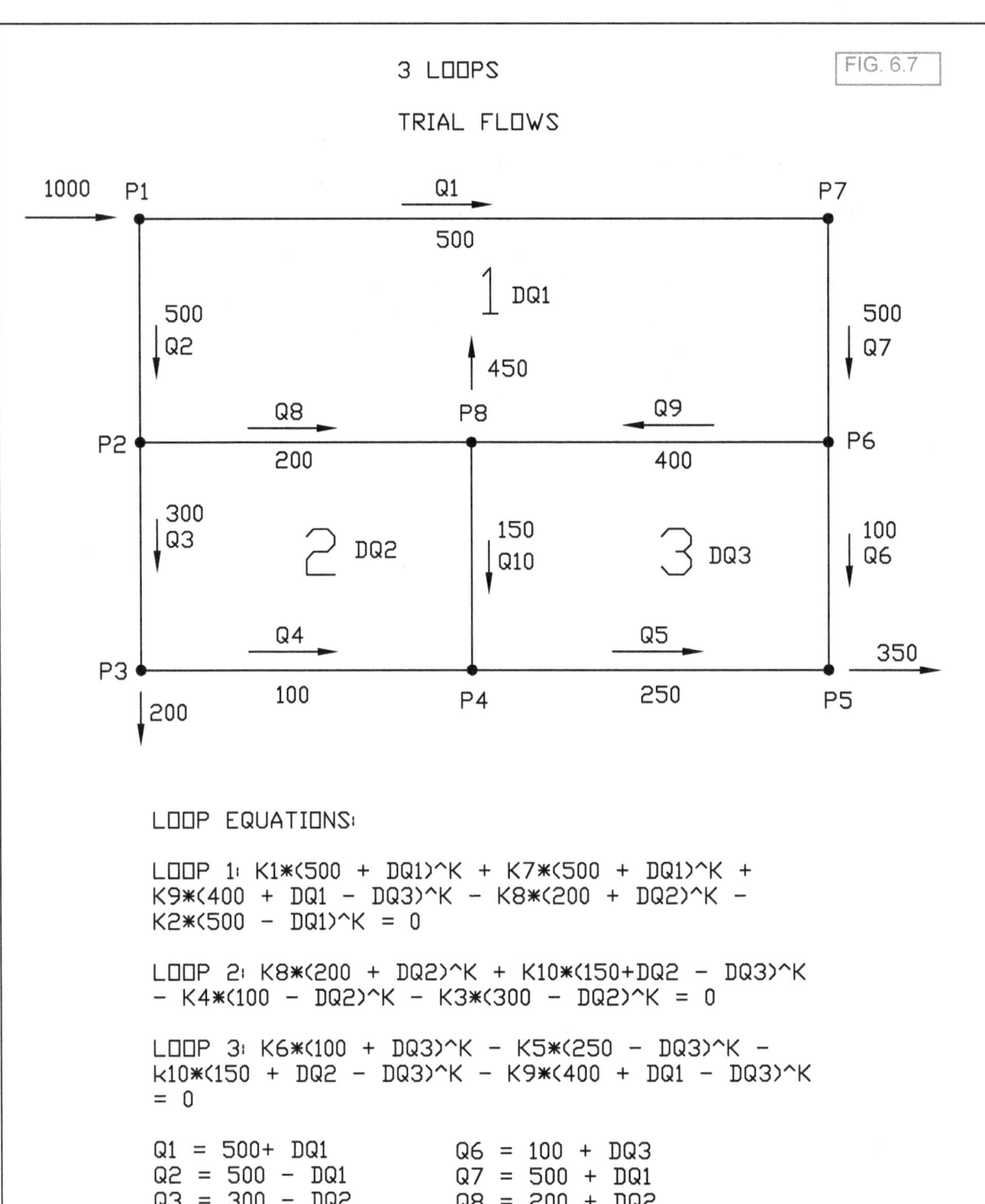

FIG. 6.7

LOOP EQUATIONS:

LOOP 1: K1*(500 + DQ1)^K + K7*(500 + DQ1)^K + K9*(400 + DQ1 - DQ3)^K - K8*(200 + DQ2)^K - K2*(500 - DQ1)^K = 0

LOOP 2: K8*(200 + DQ2)^K + K10*(150+DQ2 - DQ3)^K - K4*(100 - DQ2)^K - K3*(300 - DQ2)^K = 0

LOOP 3: K6*(100 + DQ3)^K - K5*(250 - DQ3)^K - k10*(150 + DQ2 - DQ3)^K - K9*(400 + DQ1 - DQ3)^K = 0

Q1 = 500 + DQ1   Q6 = 100 + DQ3
Q2 = 500 − DQ1   Q7 = 500 + DQ1
Q3 = 300 − DQ2   Q8 = 200 + DQ2
Q4 = 100 − DQ2   Q9 = 400 + DQ1 − DQ3
Q5 = 250 − DQ3   Q10 = 150 + DQ2 − DQ3

Rule

$K_1 = 0.002083 \cdot L_1 \cdot (100/C)^{1.85} \cdot D_1^{-4.8655}$
$K_2 = 0.002083 \cdot L_2 \cdot (100/C)^{1.85} \cdot D_2^{-4.865}$
$K_3 = 0.002083 \cdot L_3 \cdot (100/C)^{1.85} \cdot D_3^{-4.8655}$
$K_4 = 0.002083 \cdot L_4 \cdot (100/C)^{1.85} \cdot D_4^{-4.865}$
$K_5 = 0.002083 \cdot L_5 \cdot (100/C)^{1.85} \cdot D_5^{-4.8655}$
$K_6 = 0.002083 \cdot L_6 \cdot (100/C)^{1.85} \cdot D_6^{-4.865}$
$K_7 = 0.002083 \cdot L_7 \cdot (100/C)^{1.85} \cdot D_7^{-4.8655}$
$K_8 = 0.002083 \cdot L_8 \cdot (100/C)^{1.85} \cdot D_8^{-4.865}$
$K_9 = 0.002083 \cdot L_9 \cdot (100/C)^{1.85} \cdot D_9^{-4.8655}$
$K_{10} = 0.002083 \cdot L_{10} \cdot (100/C)^{1.85} \cdot D_{10}^{-4.865}$

$K_1 \cdot (500+DQ_1)^K + K_7 \cdot (500+DQ_1)^K + K_9 \cdot (400+DQ_1-DQ_3)^K - K_8 \cdot (200-DQ_1+DQ_2)^K - K_2 \cdot (500-DQ_1)^K = 0$
$K_8 \cdot (200+DQ_2)^K + K_{10} \cdot (150+DQ_2-DQ_3)^K - K_4 \cdot (100-DQ_2)^K - K_3 \cdot (300-DQ_2)^K = 0$
$K_6 \cdot (100+DQ_3)^K - K_5 \cdot (250-DQ_3)^K - K_{10} \cdot (150-DQ_3+DQ_2)^K - K_9 \cdot (400-DQ_3)^K = 0$

$Q_1 = 500 + DQ_1$
$Q_2 = 500 - DQ_1$
$Q_3 = 300 - DQ_2$
$Q_4 = 100 - DQ_3$
$Q_5 = 250 - DQ_3$
$Q_6 = 100 + DQ_3$
$Q_7 = 500 + DQ_1$
$Q_8 = 200 + DQ_2$
$Q_9 = 400 + DQ_1 - DQ_3$
$Q_{10} = 150 + DQ_2 - DQ_3$

Status	Input	Name	Output	Unit
	130	C		
		K1		
	1108	L1		
	12	D1		
		K2		
	403	L2		
	8	D2		
		K3		
	403	L3		
	6	D3		
		K4		
	504	L4		
	10	D4		
		K5		
	504	L5		
	10	D5		
		K6		
	403	L6		
	6	D6		
		K7		
	403	L7		
	8	D7		
		K8		
	504	L8		
	10	D8		
		K9		
	504	L9		
	8	D9		
		K10		
	403	L10		
	6	D10		
Guess	1	DQ1		
Guess	1	DQ3		
Guess	1	DQ2		
	1.85	K		
		Q1		
		Q2		
		Q3		
		Q4		
		Q5		
		Q6		
		Q7		
		Q8		
		Q9		
		Q10		

Status	Input	Name	Output	Unit
	130	C		
		K1	7.974023E-6	
	1108	L1		
	12	D1		
		K2	2.087692E-5	
	403	L2		
	8	D2		
		K3	8.454817E-5	
	403	L3		
	6	D3		
		K4	8.817076E-6	
	504	L4		
	10	D4		
		K5	8.806931E-6	
	504	L5		
	10	D5		
		K6	8.462395E-5	
	403	L6		
	6	D6		
		K7	2.085522E-5	
	403	L7		
	8	D7		
		K8	8.817076E-6	
	504	L8		
	10	D8		
		K9	2.608196E-5	
	504	L9		
	8	D9		
		K10	8.462395E-5	
	403	L10		
	6	D10		
		DQ1	-49.7680199	
		DQ3	120.5150419	
		DQ2	119.685628	
	1.85	K		
		Q1	450.231980	
		Q2	549.7680199	
		Q3	180.314372	
		Q4	-20.5150419	
		Q5	129.484958	
		Q6	220.5150419	
		Q7	450.231980	
		Q8	319.685628	
		Q9	229.7169382	
		Q10	149.170586	

PROBLEM 6.8 3 LOOPS, CORRECTED FLOWS

Now we know the actual direction of the flows. At this point, we introduce the orifice constant. For flow through an orifice the flow rate = k*(P)^0.5. Note that this k, is NOT the same as used in the flow equations. It varies for every opening and is used extensively in fire sprinkler design (which will be seen later). For instance, the most remote fire standpipe is required by NFPA to produce 500 gpm at a pressure of 100 psi. Then, 500 = k*(100)^0.5, so that k = 50. This concept will be used with the corrected flows to determine actual flows and pressures. Each orifice can have a different k factor. Let us assume that at P5, the flow rate = 350 gpm at 176.86' (purely arbitrary value). The pressure = 176.86/2.31 = 76.56 psi. Then 350 = k*(76.56)^0.5 and k = 350/(76.56^0.5) = 40. Then P5 = 350/40)^2. Assuming that the orifice at P8 and P3 are the same, P8 = (Q12/40)^2 and P8 = (Q11/40)^2. This is an important concept as it provides additional equations used in solving for all flows and pressures.

3 LOOPS, CORRECTED FLOWS

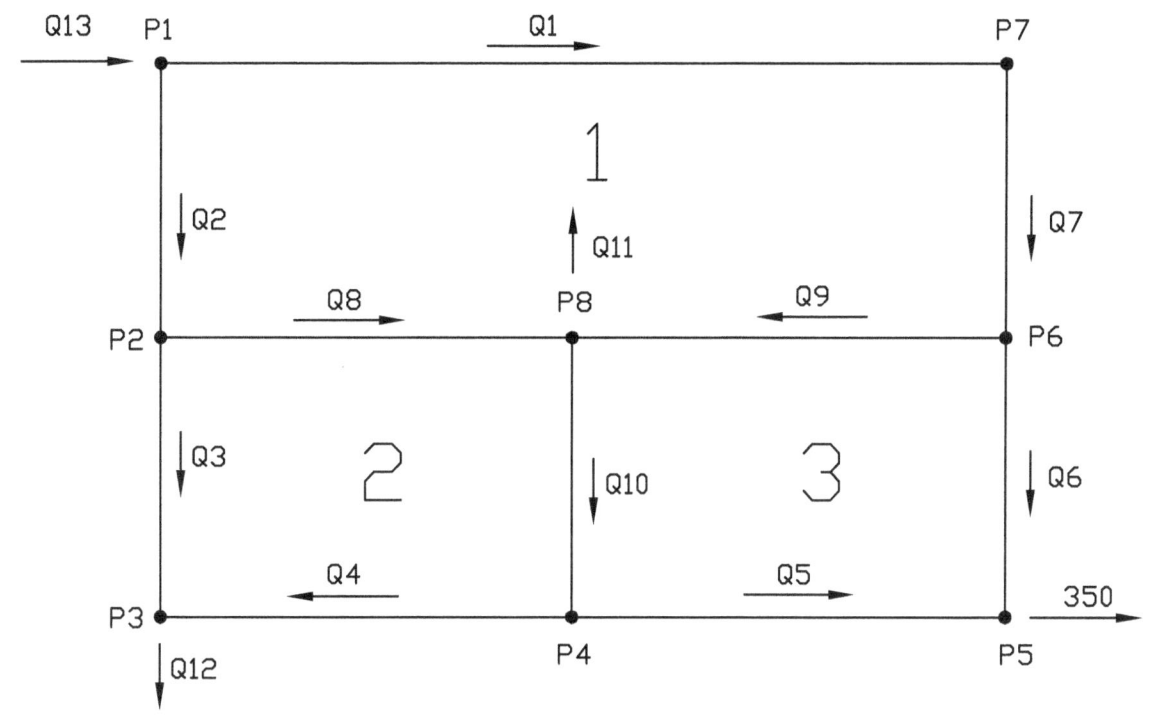

PIPE EQUATIONS:

P1 − P7 = K1*Q1^K
P1 − P2 = K2*Q2^K
P2 − P3 = K3*Q3^K
P4 − P5 = K5*Q5^K
P4 − P3 = K4*Q4^K
P6 − P5 = K6*Q6^K
P7 − P6 = K7*Q7^K
P6 − P8 = K9*Q9^K
P2 − P8 = K8*Q8^K
P8 − P4 = K10*Q10^K

NODAL EQUATIONS:

Q13 = Q1 + Q2
Q2 = Q8 + Q3
Q3 + Q4 = Q12
Q10 = Q4 + Q5
Q5 + Q6 = 350
Q7 = Q9 + Q6
Q9 + Q8 = Q11 + Q19
Q1 = Q7
P5 = 176.86
P8 = (Q11/40)^2
P3 = (Q12/40)^2

FIG. 6.8

Rule

```
K1=0.002083*L1*(100/C)^1.85*D1^-4.8655
K2=0.002083*L2*(100/C)^1.85*D2^-4.865
K3=0.002083*L3*(100/C)^1.85*D3^-4.8655
K4=0.002083*L4*(100/C)^1.85*D4^-4.865
K5=0.002083*L5*(100/C)^1.85*D5^-4.8655
K6=0.002083*L6*(100/C)^1.85*D6^-4.865
K7=0.002083*L7*(100/C)^1.85*D7^-4.8655
K8=0.002083*L8*(100/C)^1.85*D8^-4.865
K9=0.002083*L9*(100/C)^1.85*D9^-4.8655
K10=0.002083*L10*(100/C)^1.85*D10^-4.865

P1-P7=K1*Q1^K
P1-P2=K2*Q2^K
P2-P3=K3*Q3^K
P4-P3=K4*Q4^K
P4-P5=K5*Q5^K
P6-P5=K6*Q6^K
P7-P6=K7*Q7^K
P6-P8=K9*Q9^K
P2-P8=K8*Q8^K
P8-P4=K10*Q10^K

Q13=Q1+Q2
Q2=Q8+Q3
Q3+Q4=Q12
Q10=Q4+Q5
Q5+Q6=350
Q7=Q9+Q6
Q9+Q8=Q11+Q10
Q1=Q7
P8=(Q11/40)^2
P3=(Q12/40)^2
```

Status	Input	Name	Output	Unit
	130	C		
		K1	7.974023E-6	
	1108	L1		
	12	D1		
		K2	2.087692E-5	
	403	L2		
	8	D2		
		K3	8.454817E-5	
	403	L3		
	6	D3		
		K4	8.817076E-6	
	504	L4		
	10	D4		
		K5	8.806931E-6	
	504	L5		
	10	D5		
		K6	8.462395E-5	
	403	L6		
	6	D6		
		K7	2.085522E-5	
	403	L7		
	8	D7		
		K8	8.817076E-6	
	504	L8		
	10	D8		
		K9	2.608196E-5	
	504	L9		
	8	D9		
		K10	8.462395E-5	
	403	L10		
	6	D10		
	1.85	K		
Guess	1	Q1		
Guess	1	Q2		
Guess	1	Q3		
Guess	1	Q4		
Guess	1	Q5		
Guess	1	Q6		
Guess	1	Q7		
Guess	1	Q8		
Guess	1	Q9		
Guess	1	Q10		
Guess	1	P1		
Guess	1	P2		
Guess	1	P3		
Guess	1	P4		
	176.86	P5		
Guess	1	P6		
Guess	1	P7		
Guess	1	P8		

Status	Input	Name	Output	Unit
Guess	1	Q13		
Guess	1	Q12		
Guess	1	Q11		

Status	Input	Name	Output	Unit
	130	C		
		K1	7.974023E-6	
	1108	L1		
	12	D1		
		K2	2.087692E-5	
	403	L2		
	8	D2		
		K3	8.454817E-5	
	403	L3		
	6	D3		
		K4	8.817076E-6	
	504	L4		
	10	D4		
		K5	8.806931E-6	
	504	L5		
	10	D5		
		K6	8.462395E-5	
	403	L6		
	6	D6		
		K7	2.085522E-5	
	403	L7		
	8	D7		
		K8	8.817076E-6	
	504	L8		
	10	D8		
		K9	2.608196E-5	
	504	L9		
	8	D9		
		K10	8.462395E-5	
	403	L10		
	6	D10		
	1.85	K		
		Q1	633.5588889	
		Q2	783.9153174	
		Q3	309.0651808	
		Q4	222.6245007	
		Q5	34.541318019	
		Q6	315.4586818	
		Q7	633.5588889	
		Q8	474.8501367	
		Q9	318.1002077	
		Q10	257.1658189	
		P1	184.8089388	
		P2	180.0890515	
		P3	176.6719397	
		P4	176.866178	
	176.86	P5		
		P6	180.412437	
		P7	183.5928887	
		P8	179.3006276	

Status	Input	Name	Output	Unit
		Q13	1417.47420	
		Q12	531.689681	
		Q11	535.784525	

PROBLEM 6.9 4 LOOPS, 4 HYDRANTS, TRIAL FLOWS

Next, we will solve a 4-loop problem. See diagram 6.9 4LOOPS4HYDCF. The name stands for 4 loops with 4 hydrants, corrective flows. The objective is to find the input pressure to deliver a minimum flow of 500 gpm at the most remote hydrant. The solution is done in 2 parts. First, we must find the correct direction of flows. This is done with the same corrective flow technique in the previous problem.

First, we assign flows to all hydrants, starting with the most remote one, at P7 in this case. We set it for 500 gpm, our goal. The hydrants closer to the source pressure at P1, will naturally flow more, so we assign more flow to them, progressively getting larger, the closer we get to P1. Thus, Q20=520, Q18=530, and Q21=550. We really don't know what the actual flows are yet, we are just looking for the correct DIRECTION of flows. After assigning flows, we determine the flows in the pipes. They must add up to the flows assigned to the hydrants. For example the most remote hydrant at P9 = 500. We assume most is coming from P8 and guess 300 gpm. Then the rest must come from P6, or 200 gpm. The remaining flows are done like this. Again, they must add up to the total of 500+520+530+550 = 2100 gpm. Hopefully, we have guessed the correct direction. If not the corrective flow technique will do that for us.

Next, we write the loop equations, as seen on the corrective flow diagram. Loop 1, Loop 2, Loop 3 and Loop 4. Then we write the nodal equations also seen on the 2nd page of the diagram. Note that in the TK equations, we do not make a guess as to the flows. The total flow, Q17 is input only. The nodal equations will take care of the flow rates. We are looking for the corrective flows, DQ1, DQ2, DQ3 and DQ4. Once they are found the system of equations will calculate the corrected flows. Note that the solution shows that flow Q4 is a negative number. This means that the flow direction assumed is incorrect. The actual direction is from P4 to P3. This is then used to corrected the flow diagram. The link and nodal equations must reflect this correction also.

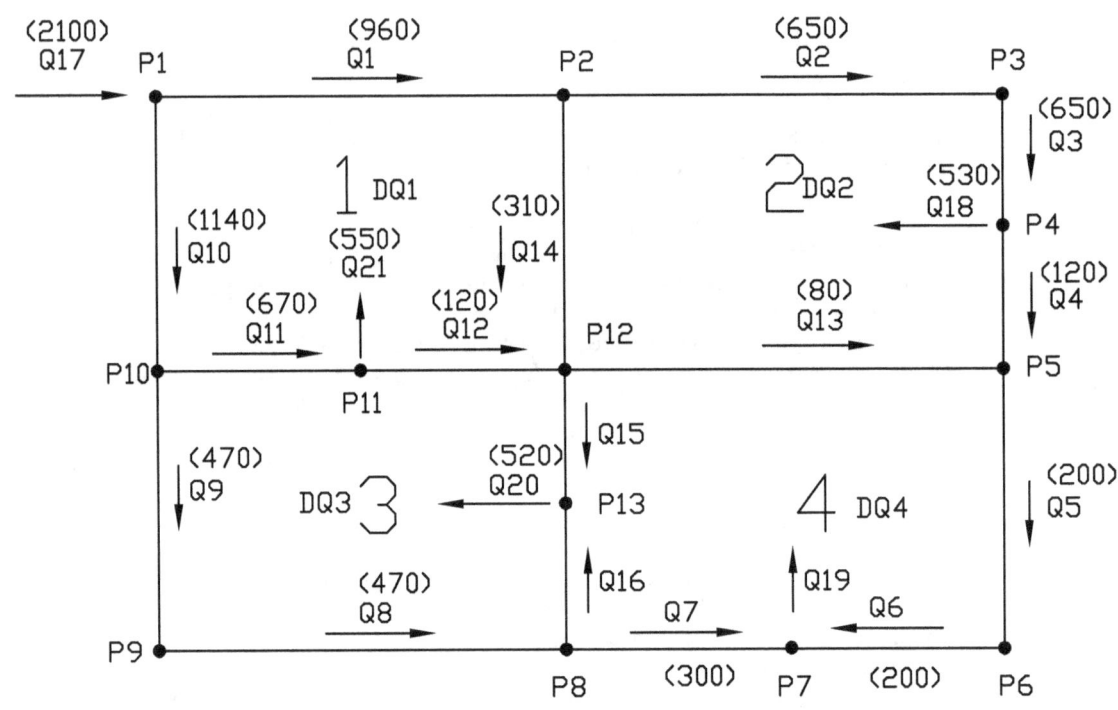

FIG. 6.9 — 4LOOPS4HYD CF

LOOP EQUATIONS:

LOOP 1:
K1*(960+DQ1)^K+K14*(310+DQ1)^K-K12*(120-DQ1+DQ3)^K
-K11*(670-DQ1+DQ3)^K-K10*(1140-DQ1)^K=0

LOOP 2:
K2*(650+DQ2)^K+K3*(650+DQ2)^K+K4*(120+DQ2)^K
-K13*(80-DQ2+DQ4)^K-K14*(310-DQ2)^K=0

LOOP 3:
K11*(670+DQ3-DQ1)^K+K12*(120+DQ3-DQ1)^K
+K15*(350+DQ3-DQ4)^K-K16*(170-DQ3+DQ4)^K
-K8*(470-DQ3)^K-K9*(470-DQ3)^K=0

LOOP 4:
K13*(80+DQ4-DQ2)^K+K5*(200+DQ4)^K+K6*(200+DQ4)^K
-K7*(300-DQ4)^K+K16*(170+DQ4-DQ3)^K
-K15*(350-DQ4+DQ3)^K=0

6.06 4 4LOOPS4HYD CF

ASSUMED FLOWS:

Q1=960
Q2=650
Q3=650
Q4=120
Q5=200
Q6=200
Q7=300
Q8=470
Q9=470
Q10=1140
Q11=670
Q12=120
Q13=80
Q14=310
Q15=350
Q16=170
Q17=2100

> Q18=530
> Q19=500 (MOST REMOTE)
> Q20=520
> Q21=550

← START WITH THESE ASSUMED FLOWS

NODAL EQUATIONS:

Q1=960+DQ1
Q2=650+DQ2
Q3=Q2
Q4=120+DQ2
Q5=200+DQ4
Q6=Q5

Q7=300−DQ4
Q8=470−DQ3
Q9=470−DQ3
Q10=1140−DQ1
Q11=670−DQ1+DQ3
Q12=120−DQ1+DQ3
Q13=80−DQ2+DQ4

Q14=310+DQ1−DQ2
Q15=350+DQ3−DQ4
Q16=170+DQ4−DQ3
Q17=Q1+Q10

Rule

$K1 = 4.52*L1*C\^-1.85*D1\^-4.8655$
$K2 = 4.52*L2*C\^-1.85*D2\^-4.8655$
$K3 = 4.52*L3*C\^-1.85*D3\^-4.8655$
$K4 = 4.52*L4*C\^-1.85*D4\^-4.8655$
$K5 = 4.52*L5*C\^-1.85*D5\^-4.8655$
$K6 = 4.52*L6*C\^-1.85*D6\^-4.8655$
$K7 = 4.52*L7*C\^-1.85*D7\^-4.8655$
$K8 = 4.52*L8*C\^-1.85*D8\^-4.8655$
$K9 = 4.52*L9*C\^-1.85*D9\^-4.8655$
$K10 = 4.52*L10*C\^-1.85*D10\^-4.8655$
$K11 = 4.52*L11*C\^-1.85*D11\^-4.8655$
$K12 = 4.52*L12*C\^-1.85*D12\^-4.8655$
$K12 = 4.52*L12*C\^-1.85*D12\^-4.8655$
$K13 = 4.52*L13*C\^-1.85*D13\^-4.8655$
$K14 = 4.52*L14*C\^-1.85*D14\^-4.8655$
$K15 = 4.52*L15*C\^-1.85*D15\^-4.8655$
$K16 = 4.52*L16*C\^-1.85*D16\^-4.8655$

K1*(960+DQ1)^K+K14*(310+DQ1)^K-K12*(120-DQ1+DQ3)^K-K11*(670-DQ1+DQ3)^K-K10*(1140-DQ1)^K=0

K2*(650+DQ2)^K+K3*(650+DQ2)^K+K4*(120+DQ2)^K-K13*(80-DQ2+DQ4)^K-K14*(310-DQ2)^K=0

K11*(670+DQ3-DQ1)^K+K12*(120+DQ3-DQ1)^K+K15*(350+DQ3-DQ4)^K-K16*(170-DQ3+DQ4)^K-K8*(470-DQ3)^K-K9*(470-l

K13*(80+DQ4-DQ3)^K+K5*(200+DQ4)^K+K6*(200+DQ4)^K-K7*(300-DQ4)^K+K16*(170+DQ4-DQ3)^K-K15*(350-DQ4+DQ3)^K

Q1=960+DQ1
Q2=650+DQ2
Q3=650+DQ2
Q4=120+DQ2
Q5=200+DQ4
Q6=Q5
Q7=300-DQ4
Q8=470-DQ3
Q9=Q8
Q10=1140-DQ1
Q11=670-DQ1+DQ3
Q12=120-DQ1+DQ3
Q13=80-DQ2+DQ4
Q14=310+DQ1-DQ2
Q15=350+DQ3-DQ4
Q16=170+DQ4-DQ3
Q17=Q1+Q10

Status	Input	Name	Output	Unit
	500	L1		
	120	C		
	6	D1		
	1.85	K		
	500	L2		
	6	D2		
	250	L3		
	6	D3		
	250	L4		
	6	D4		
	250	L5		
	4	D5		
	250	L6		
	4	D6		
	250	L7		
	4	D7		
		K1		
		K2		
		K3		
		K4		
		K5		
		K6		
		K7		
		K8		
	250	L8		
	4	D8		
		K9		
	250	L9		
	4	D9		
		K10		
	500	L10		
	6	D10		
		K11		
	250	L11		
	6	D11		
		K12		
	120	L12		
	4	D12		
		K13		
	250	L13		
	4	D13		
		K14		
	500	L14		
	6	D14		
		K15		
	250	L15		
	4	D15		
		K16		
	250	L16		
	4	D16		

Status	Input	Name	Output	Unit
Guess	1	DQ1		
Guess	1	DQ3		
Guess	1	DQ2		
Guess	1	DQ4		
		Q1		
		Q2		
		Q3		
		Q4		
		Q5		
		Q6		
		Q7		
		Q8		
		Q9		
		Q10		
		Q11		
		Q12		
		Q13		
		Q14		
		Q15		
		Q16		
	2100	Q17		

Variables

Status	Input	Name	Output	Unit	Comment
	500	L1			
	120	C			
	6	D1			
	1.85	K			
	500	L2			
	6	D2			
	250	L3			
	6	D3			
	250	L4			
	6	D4			
	250	L5			
	4	D5			
	250	L6			
	4	D6			
	250	L7			
	4	D7			
		K1	5.266574E-5		
		K2	5.266574E-5		
		K3	2.633287E-5		
		K4	2.633287E-5		
		K5	.000189352		
		K6	.000189352		
		K7	.000189352		
		K8	.000189352		
	250	L8			
	4	D8			
		K9	.000189352		
	250	L9			
	4	D9			
		K10	5.266574E-5		
	500	L10			
	6	D10			
		K11	2.633287E-5		
	250	L11			
	6	D11			
		K12	9.088903E-5		
	120	L12			

4loops4hydcf sol.tkw

Variables

Status	Input	Name	Output	Unit	Comment
	4	D12			
	250	K13	.000189352		
	4	L13			
		D13			
	500	K14	5.266574E-5		
	6	L14			
		D14			
	250	K15	.000189352		
	4	L15			
		D15			
	250	K16	.000189352		
	4	L16			
		D16			
		DQ1	116.639407!	GPM	FLOW CORRECTION, LOOP 1
		DQ3	146.254484	GPM	FLOW CORRECTION, LOOP 2
		DQ2	-93.4667214	GPM	FLOW CORRECTION, LOOP 3
		DQ4	97.6615009!	GPM	FLOW CORRECTION, LOOP 4
		Q1	1076.63940!	GPM	CORRECTED FLOW, TYPICAL
		Q2	556.533278!	GPM	
		Q3	556.533278!	GPM	
		Q4	26.5332786!	GPM	
		Q5	297.661501	GPM	
		Q6	297.661501	GPM	
		Q7	202.338499	GPM	
		Q8	323.745515!	GPM	
		Q9	323.745515!	GPM	
		Q10	1023.36059!	GPM	
		Q11	699.615076!	GPM	
		Q12	149.615076!	GPM	
		Q13	271.128222!	GPM	
		Q14	520.106128!	GPM	
		Q15	398.592983!	GPM	
		Q16	121.407016!	GPM	
	2100	Q17		GPM	TOTAL FLOW, GIVEN

PROBLEM 6.10 4 LOOPS, 4 HYDRANTS, CORRECTED FLOWS

See the diagram 4LOOPS4HYD FINAL FLOWS. Note that no flows are assigned to each pipe. We will find out what they are, using the flow, nodal and pressure equations shown on the diagram. We will be solving for 33 unknowns, including P1, the source pressure. Q19 is input as 500, minimum flow required. P1 is guessed to be about 100, but we input 50 just to see what happens. The solution shows that indeed, P1= 103.5 psi to produce 500 gpm at Q19. Just as predicted, the remaining hydrant flow rates are greater than 500 gpm. Total flow = 2128 gpm. Remember the initial guess was 2100 gpm, so we guessed pretty close.

The reader is encouraged to tweak the pipe sizes and lengths to see what happens. Be sure to save your input BEFORE solving, so you don't have to input all those guesses again. You can even input a different value of P1 to see what you get at Q19.

4LOOPS4HYD

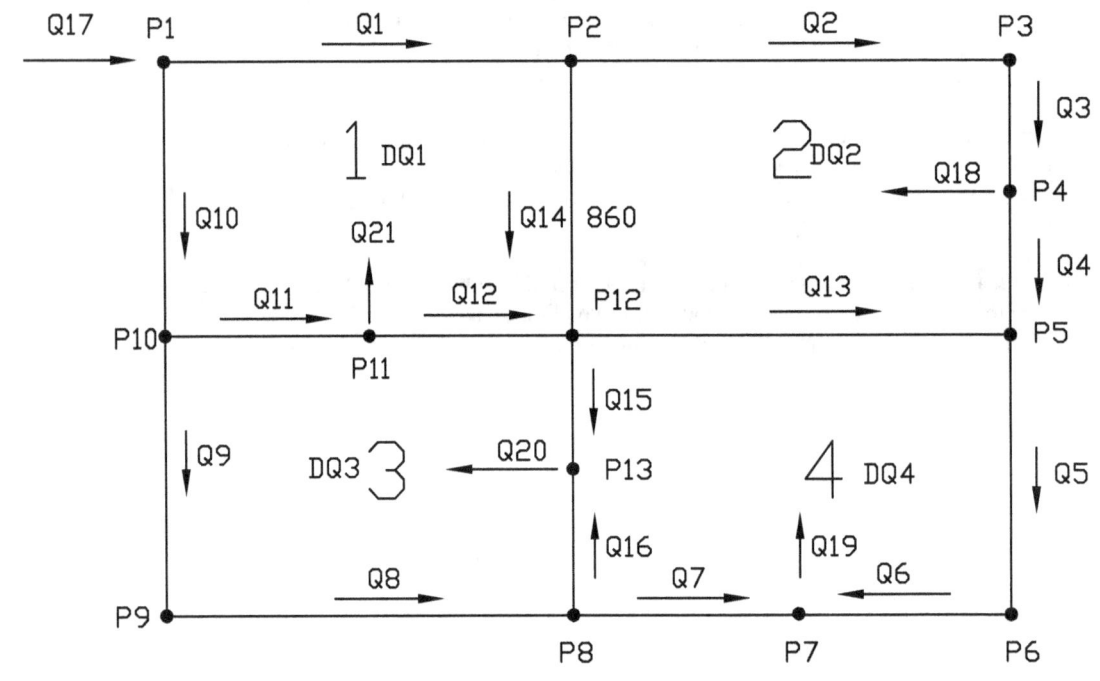

FIG. 6.10

PIPE EQUATIONS:

P1-P2=K1*Q1^K
P2-P3=K2*Q2^K
P3-P4=K3*Q3^K
P4-P5=K4*Q4^K
P5-P6=K5*Q5^K
P6-P7=K6*Q6^K
P8-P7=K7*Q7^K
P9-P8=K8*Q8^K
P10-P9=K9*Q9^K

P1-P10=K10*Q10^K
P10-P11=K11*Q11^K
P11-P12=K12*Q12^K
P12-P5=K13*Q13^K
P2-P12=K14*Q14^K
P12-P13=K15*Q15^K
P8-P13=K16*Q16^K

NODAL EQUATIONS:

Q17=Q1+Q10
Q1=Q2+Q14
Q2=Q3
Q3=Q18+Q4
Q13+Q4=Q5
Q5=Q6

Q6+Q7=Q19
Q8=Q16+Q7
Q9=Q8
Q10=Q9+Q11
Q11=Q21+Q12
Q12+Q14=Q13+Q15
Q15+Q16=Q20

P11=(Q21/64.28)^2
P3=(Q18/64.28)^2
P13=(Q29/64.28)^2
P7=(Q1/64.28)^2

6-61

Rule

$K1 = 4.52*L1*C^{-1.85}*D1^{-4.8655}$
$K2 = 4.52*L2*C^{-1.85}*D2^{-4.8655}$
$K3 = 4.52*L3*C^{-1.85}*D3^{-4.8655}$
$K4 = 4.52*L4*C^{-1.85}*D4^{-4.8655}$
$K5 = 4.52*L5*C^{-1.85}*D5^{-4.8655}$
$K6 = 4.52*L6*C^{-1.85}*D6^{-4.8655}$
$K7 = 4.52*L7*C^{-1.85}*D7^{-4.8655}$
$K8 = 4.52*L8*C^{-1.85}*D8^{-4.8655}$
$K9 = 4.52*L9*C^{-1.85}*D9^{-4.8655}$
$K10 = 4.52*L10*C^{-1.85}*D10^{-4.8655}$
$K11 = 4.52*L11*C^{-1.85}*D11^{-4.8655}$
$K12 = 4.52*L12*C^{-1.85}*D12^{-4.8655}$
$K12 = 4.52*L12*C^{-1.85}*D12^{-4.8655}$
$K13 = 4.52*L13*C^{-1.85}*D13^{-4.8655}$
$K14 = 4.52*L14*C^{-1.85}*D14^{-4.8655}$
$K15 = 4.52*L15*C^{-1.85}*D15^{-4.8655}$
$K16 = 4.52*L16*C^{-1.85}*D16^{-4.8655}$

$P1-P2 = K1*Q1^K$
$P2-P3 = K2*Q2^K$
$P3-P4 = K3*Q3^K$
$P4-P5 = K4*Q4^K$
$P5-P6 = K5*Q5^K$
$P6-P7 = K6*Q6^K$
$P8-P7 = K7*Q7^K$
$P9-P8 = K8*Q8^K$
$P10-P9 = K9*Q9^K$
$P1-P10 = K10*Q10^K$
$P10-P11 = K11*Q11^K$
$P11-P12 = K12*Q12^K$
$P12-P5 = K13*Q13^K$
$P2-P12 = K14*Q14^K$
$P12-P13 = K15*Q15^K$
$P8-P13 = K16*Q16^K$

$Q17 = Q1+Q10$
$Q1 = Q14+Q2$
$Q2 = Q3$
$Q3 = Q18+Q4$
$Q13+Q4 = Q5$
$Q5 = Q6$
$Q6+Q7 = Q19$
$Q8 = Q16+Q7$
$Q9 = Q8$
$Q10 = Q9+Q11$
$Q11 = Q21+Q12$
$Q12+Q14 = Q13+Q15$
$Q15+Q16 = Q20$

Rule
P11=(Q21/64.28)^2
P4=(Q18/64.28)^2
P13=(Q20/64.28)^2
P7=(Q19/64.28)^2

Variables

Status	Input	Name	Output	Unit	Comment
	500	L1			
	120	C			
	6	D1			
	1.85	K			
	500	L2			
	6	D2			
	250	L3			
	6	D3			
	250	L4			
	6	D4			
	250	L5			
	4	D5			
	250	L6			
	4	D6			
	250	L7			
	4	D7			
		K1			
		K2			
		K3			
		K4			
		K5			
		K6			
		K7			
		K8			
	250	L8			
	4	D8			
		K9			
	250	L9			
	4	D9			
		K10			
	500	L10			
	6	D10			
		K11			
	250	L11			
	6	D11			
		K12			
	120	L12			

Status	Input	Name	Output	Unit	Comment
	4	D12			
		K13			
	250	L13			
	4	D13			
		K14			
	500	L14			
	6	D14			
		K15			
	250	L15			
	4	D15			
		K16			
	250	L16			
	4	D16			
Guess	50	P1		PSI	INLET PRESSURE
Guess	1	P2			
Guess	1	Q1			
Guess	1	P3			
Guess	1	Q2			
Guess	1	P4			
Guess	1	Q3			
Guess	1	P5			
Guess	1	Q4			
Guess	1	P6			
Guess	1	Q5			
Guess	1	P7			
Guess	1	Q6			
Guess	1	P8			
Guess	1	Q7			
Guess	1	P9			
Guess	1	Q8			
Guess	1	P10			
Guess	1	Q9			
Guess	1	Q10			
Guess	1	P11			
Guess	1	Q11			
Guess	1	P12			
Guess	1	Q12			

Variables

Status	Input	Name	Output	Unit	Comment
Guess	1	P13			
Guess	1	Q13			
Guess	1	Q14			
Guess	1	Q15			
Guess	1	Q16			
Guess	1	Q17		GPM	TOTAL FLOW FOR ALL HYDRANTS
Guess	1	Q18		GPM	HYDRANT FLOW AT P4
Guess	500	Q19		GPM	MINIMUM FLOW AT REMOTE HYDRANT AT P7
Guess	1	Q21		GPM	HYDRANT FLOW AT P13
Guess	1	Q20		GPM	HYDRANT FLOW AT P8

Variables

4LOOPS4HYD.tkw

Variables

Status	Input	Name	Output	Unit	Comment
	500	L1			
	120	C			
	6	D1			
	1.85	K			
	500	L2			
	6	D2			
	250	L3			
	6	D3			
	250	L4			
	6	D4			
	250	L5			
	4	D5			
	250	L6			
	4	D6			
	250	L7			
	4	D7			
		K1	5.266574E-5		
		K2	5.266574E-5		
		K3	2.633287E-5		
		K4	2.633287E-5		
		K5	.000189352		
		K6	.000189352		
		K7	.000189352		
		K8	.000189352		
	250	L8			
	4	D8			
		K9	.000189352		
	250	L9			
	4	D9			
		K10	5.266574E-5		
	500	L10			
	6	D10			
		K11	2.633287E-5		
	250	L11			
	6	D11			
		K12	9.088903E-5		
	120	L12			

4LOOPS4HYD.tkw

Status	Input	Name	Output	Unit	Comment
	4	D12			
		K13	.000189352		
	250	L13			
	4	D13			
		K14	5.266574E-5		
	500	L14			
	6	D14			
		K15	.000189352		
	250	L15			
	4	D15			
		K16	.000189352		
	250	L16			
	4	D16			
		P1	103.547519	PSI	INLET PRESSURE
		P2	82.1963505		
		Q1	1074.79508		
		P3	75.5267836		
		Q2	572.965579		
		P4	72.1920001		
		Q3	572.965579		
		P5	72.1804288		
		Q4	26.8077599		
		P6	66.3425059		
		Q5	267.007312		
		P7	60.5045830		
		Q6	267.007312		
		P8	65.0434218		
		Q7	232.992687		
		P9	74.0104413		
		Q8	336.694258		
		P10	82.9774609		
		Q9	336.694258		
		Q10	1053.27480		
		P11	77.9337526		
		Q11	716.580545		
		P12	76.9796673		
		Q12	149.116396		

Variables

4LOOPS4HYD.tkw

Status	Input	Name	Output	Unit	Comment
		P13	64.0289477		
		Q13	240.199552		
		Q14	501.829503		
		Q15	410.746347		
		Q16	103.701570		
		Q17	2128.06988	GPM	TOTAL FLOW FOR ALL HYDRANTS
		Q18	546.157819	GPM	HYDRANT FLOW AT P4
	500	Q19		GPM	MINIMUM FLOW AT REMOTE HYDRANT AT P7
		Q21	567.464148	GPM	HYDRANT FLOW AT P13
		Q20	514.447918	GPM	HYDRANT FLOW AT P8

CHAP 7 - DIVIDED FLOWS

FIG. 7.1 TRADITIONAL HOUSE PIPING

FIG. 7.2 PEX HOUSE PLUMBING

This is a very common network problem, and fortunately, relatively easy to solve. There is one source of flow that goes into multiple branches. We will solve for the flows and pressures throughout the network.

Problem 7.01 is to compare a regular house plumbing system with traditional routing verses a pex piping with a manifold arrangement. See diagrams 7.01A and 7.01B. The plumbing system serves a hose bib, sink, 2 water closets, shower and lavatory. Flows for each fixture will be represented by appropriate orifice k factors. A ¾" hose bib is assumed to deliver 5 gpm @ 10 psi, so that $k=5/(10^{.5}) = 1.58$. We must convert Then, pressure at hose = $(gpm/1.58)^2$. Similarly, the sink delivers 1.5 gpm @ 5 psi, so k=0.67, the bathtub, lavatory, water closet and shower is assumed the same with k=0.67 as they are all fed with ½" pipe. Actual plumbing fixture flow rates may differ, depending on the make and model. Then, for a fixture with ½" connection, pressure = $(gpm/0.67)^2$

Darcy Weisbach and Shacham will be used for best accuracy and actual iron pipe sizes will be used. Fitting equivalent lengths are from Cameron Hydraulic Data.

Let D1=1" (1.049" ID) Let L1=10'
Let D2 = ¾" (.824" ID), Let L2 = 10' + Tee + (2) 90 Deg. Elbow = 10' + 4.2' + 2.1*2 = 18.4'
Let D3=1' (1.049"), Let L3= 8'
Let D4=1/2" (0.622", Let L4 = 10' + Tee + (2) 90 Deg. Elbows = 10' + 3.3' + 2*1.7' = 16.7'
Let D5=3/4" (.824"), Let L5 = 8'
Let D6=3/4" (.824"), Let L6 = 8' + Tee = 8' + 4.4' = 12.4'
Let D7=1/2" (.622"), Let L7 = Tee + 9' + (2) 90 Deg.Elbows = 3.3' + 9' + 2*1.7' = 15.7'
Let D8=1/2" (.622"), Let L8 = 6' + (2) 90 Deg.Elbows = 6' + 2*1.7' = 9.4'
Let D9= ¾" (.824"), Let L9 = 6' + Tee = 6 + 4.2' = 10.2'
Let D10 = ½" (.622"), Let L10 = Tee + 3' = 4.2' + 3' = 7.2'
Let D11 = 1/2" (.622"), Let L11 = 3' + (2) 90 Deg.Elbows = 3' + 2*1.7' = 6.4'
Let D12 = ½" (.622"), Let L12 = Tee + 5' + (2) 90 Deg.Elbows = 3.3' + 5 + 2*1.7' = 11.7'
Let D13= ½" (0.622"), :Let l13 = 4' + (2) 90 Elbows = 4 + 2*1.7 = 7.4'

Let P1 = 45 psig

SOLUTION: Total flow = 34.2 gpm, with all fixtures open. Note that the minimum flow (3.9) exceeds required flow. Calculation time = 0.054 sec (fast enough?). Another calculation was done to find the minimum input pressure to produce the minimum flow at the most remote fixture (1.5 gpm at P14, the WC). Minimum pressure required = 6.85 psig. Calculation time = 0.048 sec.

Next we check this against an installation using PEX piping with identical pipe sizes as iron pipe. A 1" header is used with ¾" going to the hose bib and ½" runs going to all the other fixtures.

For the PEX pipe, assume runs of 75% of the original runs, as the lines would be run diagonally to the fixtures. PEX pipe roughness factor = 0.00001.

For the hose bib at P9, L8 = ¾" Tee + .75*(10+10) = 4.2+15 = 19.2'. D2 = .824"
For the WC at P10, L9 = ½" Tee + .75*(10+8+8+8+8) = 3.3+31.5 = 34.8'
For the shower at P11, L10 = ½" Tee + .75*(10+8+8+8+9) = 3.3 + 32.25 = 35.55'
For the lav at P12, L11 = ½" Tee + .75*(10+8+8+6) = 3.3 + 24 = 27.3'
For the WC at P13, L12 = ½" Tee + .75*(10+8+8+6+3+5) = 3.3 + 30 = 33.3'
For the tub at P14, L13 = ½" Tee + .75*(10+8+9+6+4) = 3.3 + 27.75 = 31.05'
For the sink at P15, L14 = ½" Tee + .75*(10+8+10) = 3.3 + 21 = 24.3'

SOLUTION: Given minimum flow of 1.5 gpm at P13, Min pressure = 5.83 psig, total flow = 12.63 gpm. Note the difference in total flow and that almost all of the ½" connections have about 1.5 gpm. Each starts off from the main 1" header at about the same pressure. Pressure drop between takeoffs is neglible, less than 1 psi difference. Total pressure required is slightly less, about 1 psig. This shows that the PEX arrangement is feasible and is less labor intensive also.

The variety of branching systems is endless but not that hard with this method.

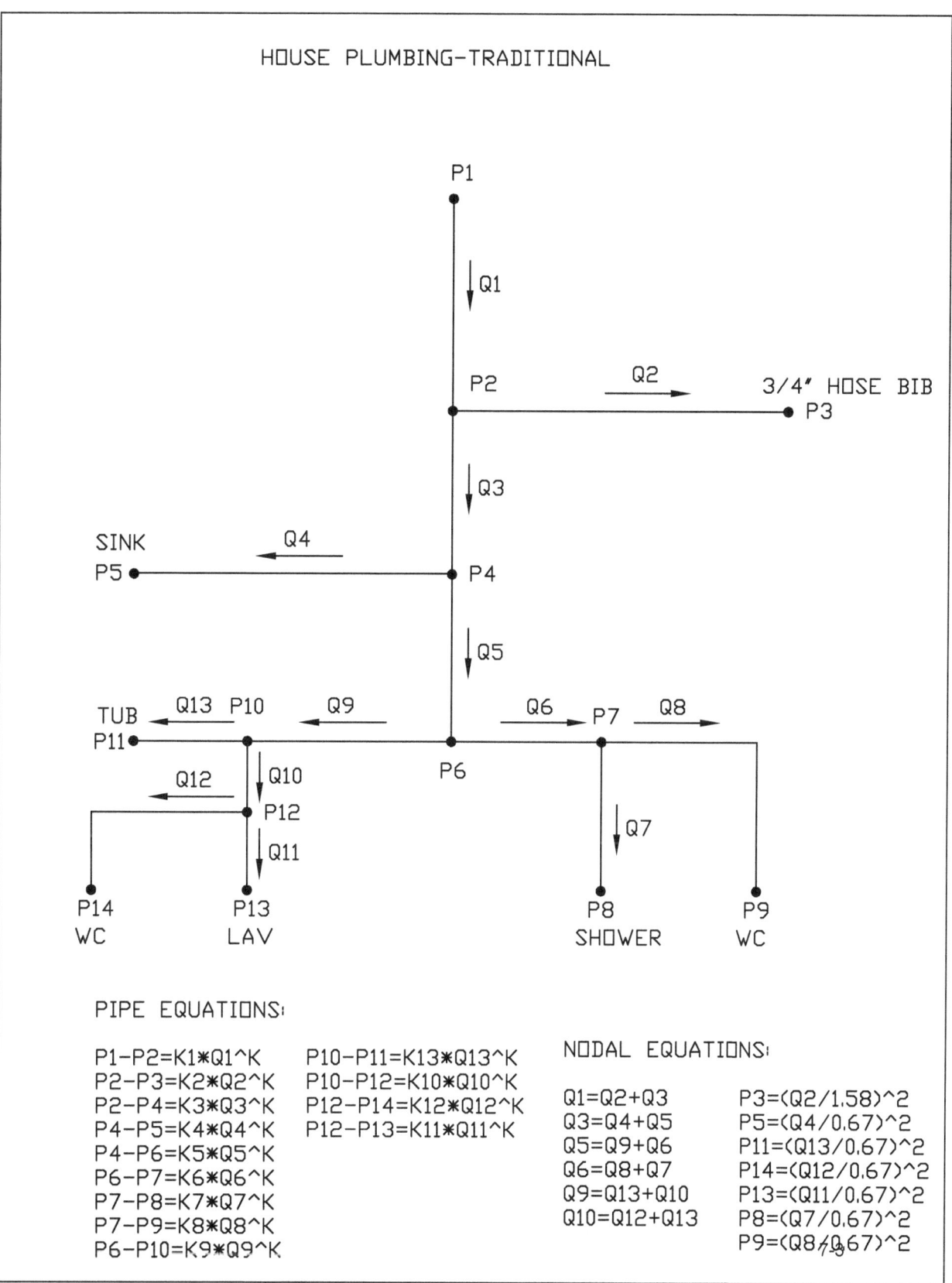

Rule

$v = a_1 + a_2*T + a_3*T^2 + a_4*T^3 + a_5*T^4 + a_6*T^5$

$rho = b_1 + b_2*T + b_3*T^2 + b_4*T^3$

R1=50.6*Q1*rho/(D1*v)
R2=50.6*Q2*rho/(D2*v)
R3=50.6*Q3*rho/(D3*v)
R4=50.6*Q4*rho/(D4*v)
R5=50.6*Q5*rho/(D5*v)
R6=50.6*Q6*rho/(D6*v)
R7=50.6*Q7*rho/(D7*v)
R8=50.6*Q8*rho/(D8*v)
R9=50.6*Q9*rho/(D9*v)
R10=50.6*Q10*rho/(D10*v)
R11=50.6*Q11*rho/(D11*v)
R12=50.6*Q12*rho/(D12*v)
R13=50.6*Q13*rho/(D13*v)

f1=(-2*log((e*12/D1)/3.7-(5.02/R1)*log((e*12/D1)/3.7+14.5/R1)))^-2
f2=(-2*log((e*12/D2)/3.7-(5.02/R2)*log((e*12/D2)/3.7+14.5/R2)))^-2
f3=(-2*log((e*12/D3)/3.7-(5.02/R3)*log((e*12/D3)/3.7+14.5/R3)))^-2
f4=(-2*log((e*12/D4)/3.7-(5.02/R4)*log((e*12/D4)/3.7+14.5/R4)))^-2
f5=(-2*log((e*12/D5)/3.7-(5.02/R5)*log((e*12/D5)/3.7+14.5/R5)))^-2
f6=(-2*log((e*12/D6)/3.7-(5.02/R6)*log((e*12/D6)/3.7+14.5/R6)))^-2
f7=(-2*log((e*12/D7)/3.7-(5.02/R7)*log((e*12/D7)/3.7+14.5/R7)))^-2
f8=(-2*log((e*12/D8)/3.7-(5.02/R8)*log((e*12/D8)/3.7+14.5/R8)))^-2
f9=(-2*log((e*12/D9)/3.7-(5.02/R9)*log((e*12/D9)/3.7+14.5/R9)))^-2
f10=(-2*log((e*12/D10)/3.7-(5.02/R10)*log((e*12/D10)/3.7+14.5/R10)))^-2
f11=(-2*log((e*12/D11)/3.7-(5.02/R11)*log((e*12/D11)/3.7+14.5/R11)))^-2
f12=(-2*log((e*12/D12)/3.7-(5.02/R12)*log((e*12/D12)/3.7+14.5/R12)))^-2
f13=(-2*log((e*12/D13)/3.7-(5.02/R13)*log((e*12/D13)/3.7+14.5/R13)))^-2

V1=.4085*Q1/(D1^2)
V2=.4085*Q2/(D2^2)
V3=.4085*Q3/(D3^2)
V4=.4085*Q4/(D4^2)
V5=.4085*Q5/(D5^2)
V6=.4085*Q6/(D6^2)
V7=.4085*Q7/(D7^2)
V8=.4085*Q8/(D8^2)
V9=.4085*Q9/(D9^2)
V10=.4085*Q10/(D10^2)
V11=.4085*Q11/(D11^2)
V12=.4085*Q12/(D12^2)
V13=.4085*Q13/(D13^2)

K1=0.01345*f1*L1/(D1^5)
K2=0.01345*f2*L2/(D2^5)

Rule

$K3 = 0.01345 \cdot f3 \cdot L3 / (D3^5)$
$K4 = 0.01345 \cdot f4 \cdot L1 / (D4^5)$
$K5 = 0.01345 \cdot f5 \cdot L5 / (D5^5)$
$K6 = 0.01345 \cdot f6 \cdot L6 / (D6^5)$
$K7 = 0.01345 \cdot f7 \cdot L7 / (D7^5)$
$K8 = 0.01345 \cdot f8 \cdot L8 / (D8^5)$
$K9 = 0.01345 \cdot f9 \cdot L9 / (D9^5)$
$K10 = 0.01345 \cdot f10 \cdot L10 / (D10^5)$
$K11 = 0.01345 \cdot f11 \cdot L11 / (D11^5)$
$K12 = 0.01345 \cdot f12 \cdot L12 / (D12^5)$
$K13 = 0.01345 \cdot f13 \cdot L13 / (D13^5)$

$P1 - P2 = K1 \cdot Q1^K$
$P2 - P3 = K2 \cdot Q2^K$
$P2 - P4 = K3 \cdot Q3^K$
$P4 - P5 = K4 \cdot Q4^K$
$P4 - P6 = K5 \cdot Q5^K$
$P6 - P7 = K6 \cdot Q6^K$
$P7 - P8 = K7 \cdot Q7^K$
$P7 - P9 = K8 \cdot Q8^K$
$P6 - P10 = K9 \cdot Q9^K$
$P10 - P11 = K13 \cdot Q13^K$
$P10 - P12 = K10 \cdot Q10^K$
$P12 - P14 = K12 \cdot Q12^K$
$P12 - P13 = K11 \cdot Q11^K$

$Q1 = Q2 + Q3$
$Q3 = Q4 + Q5$
$Q5 = Q9 + Q6$
$Q6 = Q8 + Q7$
$Q9 = Q13 + Q10$
$Q10 = Q12 + Q13$

$P3 = (Q2 / 1.58)^2$
$P5 = (Q4 / .67)^2$
$P11 = (Q13 / .67)^2$
$P14 = (Q12 / .67)^2$
$P13 = (Q11 / .67)^2$
$P8 = (Q7 / .67)^2$
$P9 = (Q8 / .67)^2$

Variables

Status	Input	Name	Output	Unit	Comment
		v			
	.000557280!	a1			
	.22430157	a2			
	78	T		Deg. F	Water temperature
	-.0091107	a3			
	.000155423!	a4			
	-1.23944E-6	a5			
	3.790416E-9	a6			
		rho			
	62.2614555	b1			
	.009339695	b2			
	-.000137041	b3			
	1.83106E-7	b4			
		R1			
Guess	1	Q1		gpm	Total flow rate
	1.049	D1		inches	ID of pipe, typical
	.00015	e			Commercial Steel roughness factor
		f1			
	10	L1			
Guess	1	V1			
		P2		psig	Pressure at node, typical
		R2			
Guess	1	Q2			
	.824	D2			
		R3			
Guess	1	Q3			
	1.049	D3			
		R4			
Guess	1	Q4			
	.622	D4			
		R5			
Guess	1	Q5			
	.824	D5			
		R6			
Guess	1	Q6			
	.824	D6			
		f2			

Status	Input	Name	Output	Unit	Comment
		f3			
		f4			
		f5			
		f6			
		V2		ft/sec	Water velocity, typical
		V3			
		V4			
		V5			
		V6			
	18.4	L2		feet	Length of pipe, typical
	8	L3			
	8	L5			
	12.4	L6			
		R7			
Guess	1	Q7			
	.622	D7			
		R8			
Guess	1	Q8			
	.622	D8			
		R9			
Guess	1	Q9			
	.824	D9			
		R10			
Guess	1	Q10			
	.622	D10			
		f7			
		f8			
		f9			
		f10			
		V7			
		V8			
		V9			
		V10			
	15.7	L7			
	9.4	L8			
	10.2	L9			
	7.2	L10			

Variables

7.01 HOUSE PLBG TRADITIONAL.tkw

Variables

Status	Input	Name	Output	Unit	Comment
		K1			
		K2			
		K3			
		K4			
		K5			
		K6			
		K7			
		K8			
		K9			
		K10			
		R11			
Guess	1	Q11			
	.622	D11			
		R12			
Guess	1	Q12			
	.622	D12			
		R13			
Guess	1	Q13			
	.622	D13			
		f11			
		f12			
		f13			
		V11			
		V12			
		V13			
		K11			
	6.4	L11			
		K12			
	11.7	L12			
		K13			
	7.4	L13			
	45	P1		psig	Inlet pressure
	2	K			
Guess	1	P3			
Guess	1	P4			
Guess	1	P5			
Guess	1	P6			

Status	Input	Name	Output	Unit	Comment
Guess	1	P7			
Guess	1	P8			
Guess	1	P9			
Guess	1	P10			
Guess	1	P11			
Guess	1	P12			
Guess	1	P14			
Guess	1	P13			

Variables

Variables

Status	Input	Name	Output	Unit	Comment
		v	.8889246838		
	.000557280	a1			
	.22430157	a2			
	78	T		Deg. F	Water temperature
	-.0091107	a3			
	.000155423	a4			
	-1.23944E-6	a5			
	3.790416E-9	a6			
		rho	62.2430875		
	62.2614555	b1			
	.009339695	b2			
	-.000137041	b3			
	1.83106E-7	b4			
		R1	115659.735		
		Q1	34.2437251	gpm	Flow rate
	1.049	D1		inches	ID of pipe, typical
	.00015	e			Commercial Steel roughness factor
		f1	.024126670		
	10	L1			
		V1	12.7122401		
		P2	42.0042706	psig	Pressure at node, typical
		R2	43083.9243		
		Q2	10.0199553		
	.824	D2			
		R3	81816.8813		
		Q3	24.2237697		
		D3			
	1.049	R4	24132.6726		
		Q4	4.23661624		
		D4			
	.622	R5	85941.0025		
		Q5	19.9871535		
		D5			
	.824	R6	34644.4915		
		Q6	8.05721077		
		D6			
	.824	f2	.027318246		

Status	Input	Name	Output	Unit	Comment
		f3	.024711789		
		f4	.030540334		
		f5	.025813802		
		f6	.027973770		
		V2	6.02841893	ft/sec	Water velocity, typical
		V3	8.99254904		
		V4	4.47332466		
		V5	12.0250969		
		V6	4.84755072		
	18.4	L2		feet	Length of pipe, typical
	8	L3			
	8	L5			
	12.4	L6			
		R7	22877.4146		
	.622	Q7	4.01624917		
		D7			
		R8	23018.1818		
	.622	Q8	4.04096159		
		D8			
		R9	51296.5110		
		Q9	11.9299427		
	.824	D9			
		R10	45074.4969		
	.622	Q10	7.91306247		
		D10			
		f7	.030743748		
		f8	.030720020		
		f9	.026861624		
		f10	.028647635		
		V7	4.24064522		
		V8	4.26673838		
		V9	7.17754621		
		V10	8.35518145		
	15.7	L7			
	9.4	L8			
	10.2	L9			
	7.2	L10			

Variables

Status	Input	Name	Output	Unit	Comment
		K1	.002554713		
		K2	.017797412		
		K3	.002093336		
		K4	.044120878		
		K5	.007311865		
		K6	.012281712		
		K7	.069731151		
		K8	.041717638		
		K9	.009701048		
		K10	.029798310		
		R11	22309.4938		
	.622	Q11	3.916547772		
		D11			
		R12	22193.4873		
	.622	Q12	3.89618218		
		D12			
		R13	22881.0096		
	.622	Q13	4.01688029		
		D13			
		f11	.030841996		
		f12	.030862577		
		f13	.030743139		
		V11	4.135373250		
		V12	4.11386984		
		V13	4.241311600		
		K11	.028516276		
	6.4	L11			
		K12	.052166105		
	11.7	L12			
		K13	.032866261		
	7.4	L13			
	45	P1		psig	Inlet pressure
	2	K			
		P3	40.2174304		
		P4	40.7759229		
		P5	39.98400455		
		P6	37.85493915		

Variables

7.01 house plbg traditional sol.tkw

Status	Input	Name	Output	Unit	Comment
		P7	37.0576271		
		P8	35.9328459		
		P9	36.3764044		
		P10	36.4742564		
		P11	35.9439510		
		P12	34.6083939		
		P14	33.8165005		
		P13	34.1709728		

Variables

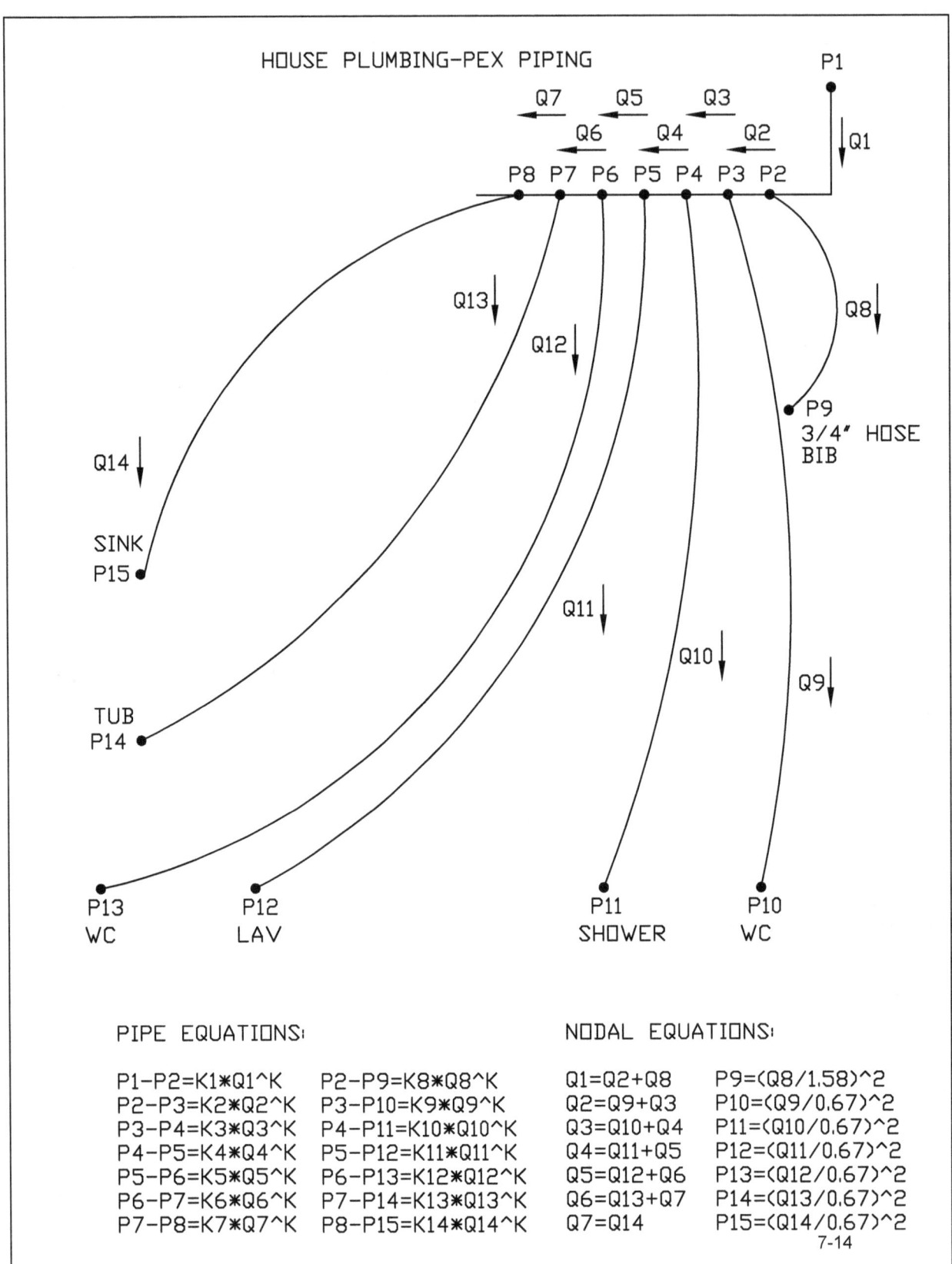

Status Rule

Satisfi v=a1+a2*T+a3*T^2+a4*T^3+a5*T^4+a6*T^5

Satisfi rho=b1+b2*T+b3*T^2+b4*T^3

Satisfi R1=50.6*Q1*rho/(D1*v)
Satisfi R2=50.6*Q2*rho/(D2*v)
Satisfi R3=50.6*Q3*rho/(D3*v)
Satisfi R4=50.6*Q4*rho/(D4*v)
Satisfi R5=50.6*Q5*rho/(D5*v)
Satisfi R6=50.6*Q6*rho/(D6*v)
Satisfi R7=50.6*Q7*rho/(D7*v)
Satisfi R8=50.6*Q8*rho/(D8*v)
Satisfi R9=50.6*Q9*rho/(D9*v)
Satisfi R10=50.6*Q10*rho/(D10*v)
Satisfi R11=50.6*Q11*rho/(D11*v)
Satisfi R12=50.6*Q12*rho/(D12*v)
Satisfi R13=50.6*Q13*rho/(D13*v)
Satisfi R14=50.6*Q14*rho/(D14*v)

Satisfi f1=(-2*log((e*12/D1)/3.7-(5.02/R1)*log((e*12/D1)/3.7+14.5/R1)))^-2
Satisfi f2=(-2*log((e*12/D2)/3.7-(5.02/R2)*log((e*12/D2)/3.7+14.5/R2)))^-2
Satisfi f3=(-2*log((e*12/D3)/3.7-(5.02/R3)*log((e*12/D3)/3.7+14.5/R3)))^-2
Satisfi f4=(-2*log((e*12/D4)/3.7-(5.02/R4)*log((e*12/D4)/3.7+14.5/R4)))^-2
Satisfi f5=(-2*log((e*12/D5)/3.7-(5.02/R5)*log((e*12/D5)/3.7+14.5/R5)))^-2
Satisfi f6=(-2*log((e*12/D6)/3.7-(5.02/R6)*log((e*12/D6)/3.7+14.5/R6)))^-2
Satisfi f7=(-2*log((e*12/D7)/3.7-(5.02/R7)*log((e*12/D7)/3.7+14.5/R7)))^-2
Satisfi f8=(-2*log((e*12/D8)/3.7-(5.02/R8)*log((e*12/D8)/3.7+14.5/R8)))^-2
Satisfi f9=(-2*log((e*12/D9)/3.7-(5.02/R9)*log((e*12/D9)/3.7+14.5/R9)))^-2
Satisfi f10=(-2*log((e*12/D10)/3.7-(5.02/R10)*log((e*12/D10)/3.7+14.5/R10)))^-2
Satisfi f11=(-2*log((e*12/D11)/3.7-(5.02/R11)*log((e*12/D11)/3.7+14.5/R11)))^-2
Satisfi f12=(-2*log((e*12/D12)/3.7-(5.02/R12)*log((e*12/D12)/3.7+14.5/R12)))^-2
Satisfi f13=(-2*log((e*12/D13)/3.7-(5.02/R13)*log((e*12/D13)/3.7+14.5/R13)))^-2
Satisfi f14=(-2*log((e*12/D14)/3.7-(5.02/R14)*log((e*12/D14)/3.7+14.5/R14)))^-2

Satisfi V1= 4085*Q1/(D1^2)
Satisfi V2= 4085*Q2/(D2^2)
Satisfi V3= 4085*Q3/(D3^2)

Rules   7.01 house plbg pex piping min flow sol.tkw

Status Rule

Satisfi V4=.4085*Q4/(D4^2)
Satisfi V5=.4085*Q5/(D5^2)
Satisfi V6=.4085*Q6/(D6^2)
Satisfi V7=.4085*Q7/(D7^2)
Satisfi V8=.4085*Q8/(D8^2)
Satisfi V9=.4085*Q9/(D9^2)
Satisfi V10=.4085*Q10/(D10^2)
Satisfi V11=.4085*Q11/(D11^2)
Satisfi V12=.4085*Q12/(D12^2)
Satisfi V13=.4085*Q13/(D13^2)
Satisfi V14=.4085*Q14/(D14^2)

Satisfi K1=0.01345*f1*L1/(D1^5)
Satisfi K2=0.01345*f2*L2/(D2^5)
Satisfi K3=0.01345*f3*L3/(D3^5)
Satisfi K4=0.01345*f4*L1/(D4^5)
Satisfi K5=0.01345*f5*L5/(D5^5)
Satisfi K6=0.01345*f6*L6/(D6^5)
Satisfi K7=0.01345*f7*L7/(D7^5)
Satisfi K8=0.01345*f8*L8/(D8^5)
Satisfi K9=0.01345*f9*L9/(D9^5)
Satisfi K10=0.01345*f10*L10/(D10^5)
Satisfi K11=0.01345*f11*L11/(D11^5)
Satisfi K12=0.01345*f12*L12/(D12^5)
Satisfi K13=0.01345*f13*L13/(D13^5)
Satisfi K14=0.01345*f14*L14/(D14^5)

Satisfi P1-P2=K1*Q1^K
Satisfi P2-P3=K2*Q2^K
Satisfi P3-P4=K3*Q3^K
Satisfi P4-P5=K4*Q4^K
Satisfi P5-P6=K5*Q5^K
Satisfi P6-P7=K6*Q6^K
Satisfi P7-P8=K7*Q7^K
Satisfi P2-P9=K8*Q8^K
Satisfi P3-P10=K9*Q9^K
Satisfi P4-P11=K10*Q10^K

Status	Rule
Satisfi	P5-P12=K11*Q11^K
Satisfi	P6-P13=K12*Q12^K
Satisfi	P7-P14=K13*Q13^K
Satisfi	P8-P15=K14*Q14^K
Satisfi	Q1=Q2+Q8
Satisfi	Q2=Q9+Q3
Satisfi	Q3=Q10+Q4
Satisfi	Q4=Q11+Q5
Satisfi	Q5=Q12+Q6
Satisfi	Q6=Q13+Q7
Satisfi	Q7=Q14
Satisfi	P9=(Q8/1.58)^2
Satisfi	P10=(Q9/.67)^2
Satisfi	P11=(Q10/.67)^2
Satisfi	P12=(Q11/.67)^2
Satisfi	P13=(Q12/.67)^2
Satisfi	P14=(Q13/.67)^2
Satisfi	P15=(Q14/.67)^2

Status	Input	Name	Output	Unit	Comment
		v			
	.000557280	a1			
	.22430157	a2			
	78	T			
	-.0091107	a3			
	.000155423	a4			
	-1.23944E-6	a5			
	3.790416E-9	a6			
		rho			
	62.2614555	b1			
	.009339695	b2			
	-.000137041	b3			
	1.83106E-7	b4			
		R1			
Guess	1	Q1			
	1.049	D1			
	.0005	e			
		f1			
	10	L1			
		V1			
Guess	1	P2			
		R2			
Guess	1	Q2			
	1.049	D2			
		R3			
Guess	1	Q3			
	1.049	D3			
		R4			
Guess	1	Q4			
	1.049	D4			
		R5			
Guess	1	Q5			
	1.049	D5			
		R6			
Guess	1	Q6			
	1.049	D6			
		f2			

Variables

Status	Input	Name	Output	Unit	Comment
		f3			
		f4			
		f5			
		f6			
		V2			
		V3			
		V4			
		V5			
		V6			
	.5	L2			
	.5	L3			
	.5	L5			
	.5	L6			
		R7			
Guess	1	Q7			
	1.049	D7			
		R8			
Guess	1	Q8			
	.824	D8			
		R9			
Guess	1	Q9			
	.622	D9			
		R10			
Guess	1	Q10			
	.622	D10			
		f7			
		f8			
		f9			
		f10			
		V7			
		V8			
		V9			
		V10			
	24.3	L7			
	19.7	L8			
	34.8	L9			
	35.55	L10			

Variables

HOUSE PLBG PEX PIPING.tkw

Variables

Status	Input	Name	Output	Unit	Comment
		K1			
		K2			
		K3			
		K4			
		K5			
		K6			
		K7			
		K8			
		K9			
		K10			
		R11			
Guess	1	Q11			
	.622	D11			
		R12			
Guess	1	Q12			
	.622	D12			
		R13			
Guess	1	Q13			
	.622	D13			
		f11			
		f12			
		f13			
		V11			
		V12			
		V13			
		K11			
	27.3	L11			
		K12			
	33.3	L12			
		K13			
	31.05	L13			
		R14			
Guess	1	Q14			
	.622	D14			
		f14			
		V14			
		K14			

HOUSE PLBG PEX PIPING.tkw

Variables

Status	Input	Name	Output	Unit	Comment
	24.3	L14			
	103.95	P1			
	2	K			
Guess	1	P3			
Guess	1	P4			
Guess	1	P5			
Guess	1	P6			
Guess	1	P7			
Guess	1	P8			
Guess	1	P9			
Guess	1	P10			
Guess	1	P11			
Guess	1	P12			
Guess	1	P13			
Guess	1	P14			
Guess	1	P15			

Variables

Status	Input	Name	Output	Unit	Comment
		v	.8889246839		
	.000557280	a1			
	.22430157	a2			
	78	T		Deg.F	Water temperature
	-.0091107	a3			
	.000155423	a4			
	-1.23944E-6	a5			
	3.790416E-9	a6			
		rho	62.2430875		
	62.2614555	b1			
	.009339695	b2			
	-.000137041	b3			
	1.83106E-7	b4			
		R1	42672.6056		Reynolds number, typical
		Q1	12.6342065	GPM	Total flow rate
	1.049	D1		INCHES	Pipe ID, typical
	.00001	e			Plastic pipe roughness factor
		f1	.022118265		Friction factor, typical
	10	L1		Feet	Pipe length, typical
		V1	4.69017509	Ft/Sec	Flow velocity, typical
		P2	5.46152758	psig	Pressure at node, typical
		R2	30493.4543		
		Q2	9.02828862		
	1.049	D2			
		R3	25392.9322		
		Q3	7.51816172		
	1.049	D3			
		R4	20299.6819		
		Q4	6.01018781		
	1.049	D4			
		R5	15212.2269		
		Q5	4.50392971		
	1.049	D5			
		R6	10156.5103		
		Q6	3.00706851		
	1.049	D6			
		f2	.023792729		

Status	Input	Name	Output	Unit	Comment
		f3	.024792254		
		f4	.026106517		
		f5	.027963217		
		f6	.030911991		
		V2	3.35155629		
		V3	2.79095444		
		V4	2.23115184		
		V5	1.67198620		
		V6	1.11630895		
	.5	L2			
	.5	L3			
	.5	L5			
	.5	L6			
		R7	5090.19228		
		Q7	1.50706851		
	1.049	D7			
		R8	15504.769		
		Q8	3.60591789	gpm	Flow @ hose bib
	.824	D8			
		R9	8602.00590		
		Q9	1.51012689	gpm	Flow @ WC
	.622	D9			
		R10	8589.74204		
		Q10	1.50797391	gpm	Flow @ shower
	.622	D10			
		f7	.037103311		
		f8	.027892306		
		f9	.032368005		
		f10	.032379855		
		V7	.559466494		
		V8	2.16946910		
		V9	1.59450077		
		V10	1.59222749		
	24.3	L7			
	19.7	L8			
	34.8	L9			
	35.55	L10			

Variables

7.01 house plbg pex piping min flow sol.tkw

Variables

Status	Input	Name	Output	Unit	Comment
		K1	.0023420480		
		K2	.0001125967		
		K3	.000131259!		
		K4	.0027643544		
		K5	.0001480478		
		K6	.0001163659		
		K7	.0095469311		
		K8	.019455253		
		K9	.162729219		
		K10	.166297172!		
		R11	8579.968418		
	.622	Q11	1.5062581	gpm	Flow @ lav
		D11			
		R12	8526.44165!		
	.622	Q12	1.49686119!	gpm	Flow @ WC
		D12			
		R13	8544.32094		
	1.5	Q13		gpm	Flow @ bathtub
	.622	D13			
		f11	.032389315!		
		f12	.032441390!		
		f13	.032423946!		
		V11	1.59041581!		
		V12	1.58049389!		
		V13	1.58380806!		
		K11	.127742312!		
	27.3	L11			
		K12	.156068069!		
	33.3	L12			
		K13	.145444678!		
	31.05	L13			
		R14	8584.58473!		
	.622	Q14	1.507068511	gpm	Flow @ sink
		D14			
		f14	.032384845!		
		V14	1.59127151!		
		K14	.113689003!		

Status	Input	Name	Output	Unit	Comment
	24.3	L14			
		P1	5.83537282	psig	Minimum pressure at inlet
	2	K			
		P3	5.45125995		
		P4	5.44384080		
		P5	5.34398579		
		P6	5.34098258		
		P7	5.33950269		
		P8	5.31781917		
		P9	5.20855785		
		P10	5.0801587		
		P11	5.06568347		
		P12	5.05416231		
		P13	4.99129749		
		P14	5.01225217		
		P15	5.05960240		

Variables

7.01 house plbg pex piping min flow sol.tkw

CHAPTER 8 - FIRE PROTECTION

PROBLEM 8.1 12 HEAD TREE SYSTEM

Fire Sprinkler design is all about piping networks, "Tree" style, grid, or hybrids of the tree + grid. As per NFPA 13, we will use a minimum pipe size of 1". NFPA13R allows ¾" but we will not go there. NFPA uses the Hazen Williams equation, so it will be used here also.

First, consider the short branch off a fire main that feeds a sprinkler head. It is called a "Sprig", probably because it resembles a sprig off a tree limb. The sprig is of course 1" in size and about 6" to 12" long. Some software out there just models the sprinkler head right on the branch, ignoring the sprig. We will not. There are 2 ways to account for it. Model the sprig as another branch, accounting for the loss through the 1" Tee + the short length just upstream of the head. That will work of course, but adds more equations to solve. The other way is to modify the sprinkler k-factor to account for the piping loss just before the head. A spreadsheet was created to do just that. The sprig length was set to 1', C-factor = 120 and a standard ½" head with k-factor of 5.6 was used. The flow was varied from the minimum 14.8 (@ 7.0 psi per NFPA 13) up to 28 gpm in 2 gpm increments. The total loss due to the sprig length + sprinkler head was calculated. The total loss, psi = (gpm/equiv k-factor)^2. (Where have you seen that before?). Equivalent k-factor = gpm/(dp^.5). The equivalent k-factor does not vary much, from 5.43 to 5.44. Being conservative, we will use 5.44 as the equivalent k-factor. See the page with the sprig equivalent k-factor spreadsheet.

Another consideration is elevation. The sprinkler system is an open system when it flows. That is, open to atmospheric pressure. Thus the elevation of the sprinkler heads must be accounted for because that is a loss also, just as is friction. We will add the elevation to each node as we go from the most remote head to the base of the riser. In some cases, the most remote head is NOT the one most distant (horizontally) from the riser. It might be the 1st head if its elevation is high enough. In order to be consistant with psi calculation, elevation heads are converted to psi. Remember that 2.31' = 1 psi, so divide the elevation in feet by 2.31 to obtain the elevation in psi. Actually the only elevations that are important is at the point of discharge, so the head elevation will be added to the required pressure.

As mentioned before, NFPA usually calls for a minimum of 7.0 psi at the most remote head. In some cases, like high-piled storage, a minimum of 45 or 50 psi is required as that is what the UL listing of the head calls for. We will just stick to 7.0 psi for our purposes since this book is not about just fire protection. Note that the remote head might require more than 7.0 psi in order to deliver the required coverage of the head.

We will consider a good old "Tree" style layout, as that is the most common. Assume the building is an office building, light hazard, with a minimum requirement of 0.1 gpm/sq.ft. over the most remote 1500 sq.ft.

See Diagram 8.1 TREE STYLE FIRE SPRINKLER SYSTEM. This was solved with TK. Schedule 10 pipe was used to provide maximum flow capacity. This piping is commonly used in fire sprinkler service.

The resulting minimum flow from the most remote head at P3 = 12.4 gpm with a residual pressure = 10.38 psi which is 7.0 + elevation pressure. Total calculated flow = 214.7 gpm with a required pressure = 31.5 psi. Note that many velocities are in excess of 10 ft/sec. If the required pressure is in excess of what is available, the pipe sizes with velocities > 10 ft./sec. could be enlarged to the next size up. Try it and see what the resulting pressure required will be. This is a good exercise for the reader.

Note that Darcy-Weisbach and the Shacham equation could be used also.

Now, commercial programs can do this and more, but this technique shows the actual equations and how it is done. If one checks fire sprinkler design only rarely, why spend the money on a commercial program when this is available? It also helps one to really understand the hydraulics.

SPRIG EQUIVALENT K-FACTOR								
PIPE C-FACTOR	120	120	120	120	120	120	120	120
SPRIG SIZE (INCHES)	1	1	1	1	1	1	1	1
SPRIG ID (INCHES)	1.049	1.049	1.049	1.049	1.049	1.049	1.049	1.049
SPRIG LENGTH (FEET)	1	1	1	1	1	1	1	1
FITTING EQ. LENGTH (FEET)	5	5	5	5	5	5	5	5
TOTAL EQ. LENGTH (FEET)	6	6	6	6	6	6	6	6
FLOW RATE (GPM)	14.8	16	18	20	22	24	26	28
HEAD K-FACTOR	5.6	5.6	5.6	5.6	5.6	5.6	5.6	5.6
DP FOR SPRIG (PSI)	0.45	0.52	0.64	0.78	0.93	1.09	1.27	1.46
DP FOR HEAD (PSI)	6.98	8.16	10.33	12.76	15.43	18.37	21.56	25.00
TOTAL DP (PSI)	7.43	8.68	10.97	13.54	16.37	19.46	22.82	26.46
EQUIV. K-FACTOR FOR SPRIG AND HEAD	5.43	5.43	5.43	5.44	5.44	5.44	5.44	5.44

12 HEAD TREE

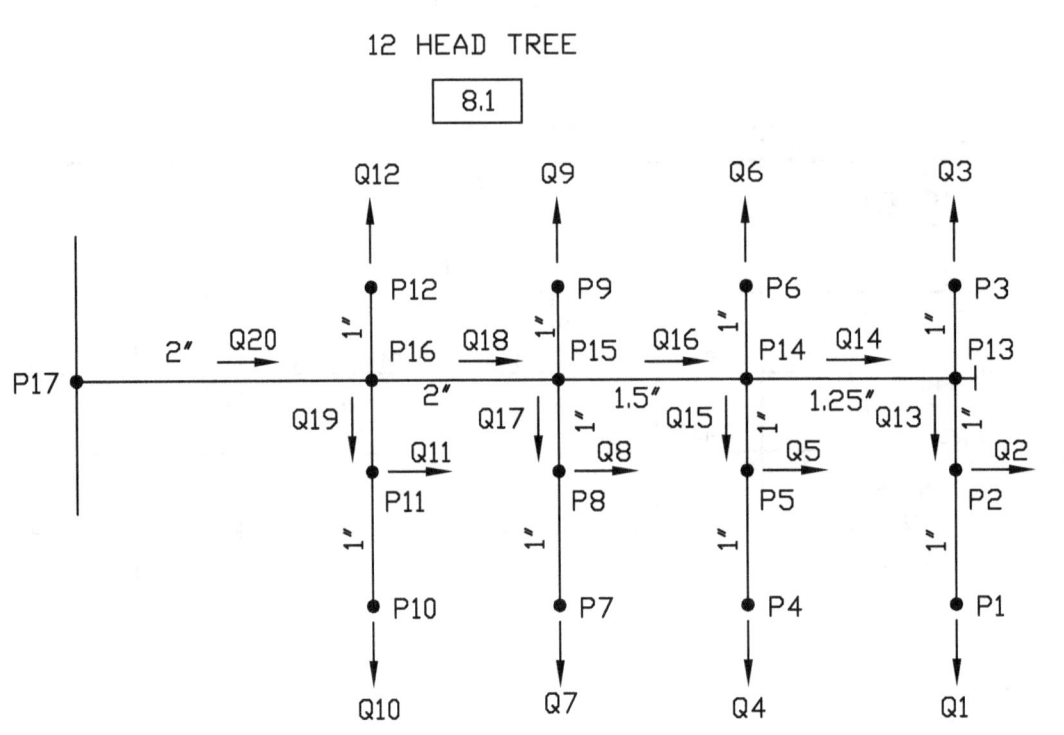

8.1

PIPE EQUATIONS:

P17−P16=K20∗Q20^K
P16−P15=K18∗Q18^K
P15−P14=K16∗Q16^K
P14−P13=K14∗Q14^K
P16−P12=K12∗Q12^K
P16−P11=K19∗Q19^K
P11−P10=K10^Q10^K
P15−P9=K9∗Q9^K
P15−P8=K17∗Q17^K
P8−P7=K7∗Q7^K
P14−P6=K6∗Q6^K
P14−P5=K15∗Q15^K
P5−P4=K4∗Q4^K
P13−P3=K3∗Q3^K
P13−P2=K13∗Q13^K
P2−P1=K1∗Q1^K

NODAL EQUATIONS:

Q1+Q2=Q13
Q14=Q3+Q13
Q15=Q5+Q4
Q16=Q6+Q15+Q14
Q17=Q8+Q7
Q18=Q9+Q17+Q16
Q19=Q11+Q10
Q20=Q12+Q19+Q18

PRESSURE EQUATIONS:

P1=(Q1/KS)^2+HE
P2=(Q2/KS)^2+HE
P3=(Q3/KS)^2+HE
P4=(Q4/KS)^2+HE
P5=(Q5/KS)^2+HE
P6=(Q6/KS)^2+HE
P7=(Q7/KS)^2+HE
P8=(Q8/KS)^2+HE
P9=(Q9/KS)^2+HE
P10=(Q10/KS)^2+HE
P11=(Q11/KS)^2+HE
P12=(Q12/KS)^2+HE

Rule

;TK SOLVER, 12 HEAD FIRE SPRINKLER SYSTEM

;COMPUTE PIPE CONSTANTS, HAZEN WILLIAMS EQUATION
K1=4.52*L1*C^-1.85*D1^-4.8655
K3=4.52*L3*C^-1.85*D3^-4.8655
K4=4.52*L4*C^-1.85*D4^-4.8655
K6=4.52*L6*C^-1.85*D6^-4.8655
K7=4.52*L7*C^-1.85*D7^-4.8655
K9=4.52*L9*C^-1.85*D9^-4.8655
K10=4.52*L10*C^-1.85*D10^-4.8655
K12=4.52*L12*C^-1.85*D12^-4.8655
K13=4.52*L13*C^-1.85*D13^-4.8655
K14=4.52*L14*C^-1.85*D14^-4.8655
K15=4.52*L15*C^-1.85*D15^-4.8655
K16=4.52*L16*C^-1.85*D16^-4.8655
K17=4.52*L17*C^-1.85*D17^-4.8655
K18=4.52*L18*C^-1.85*D18^-4.8655
K19=4.52*L19*C^-1.85*D19^-4.8655
K20=4.52*L20*C^-1.85*D20^-4.8655

;COMPUTE FLOW VELOCITIES
V1=0.4085*Q1/(D1^2)
V3=0.4085*Q3/(D3^2)
V4=0.4085*Q4/(D4^2)
V6=0.4085*Q6/(D6^2)
V7=0.4085*Q7/(D7^2)
V9=0.4085*Q9/(D9^2)
V10=0.4085*Q10/(D10^2)
V12=0.4085*Q12/(D12^2)
V13=0.4085*Q13/(D13^2)
V14=0.4085*Q14/(D14^2)
V15=0.4085*Q15/(D15^2)
V16=0.4085*Q16/(D16^2)
V17=0.4085*Q17/(D17^2)
V18=0.4085*Q18/(D18^2)

Rule

$V19 = 0.4085 \cdot Q19/(D19^2)$

$V20 = 0.4085 \cdot Q20/(D20^2)$

;PIPE FLOW EQUATIONS

$P17 - P16 = K20 \cdot Q20^K$

$P16 - P15 = K18 \cdot Q18^K$

$P15 - P14 = K16 \cdot Q16^K$

$P14 - P13 = K14 \cdot Q14^K$

$P16 - P12 = K12 \cdot Q12^K$

$P16 - P11 = K19 \cdot Q19^K$

$P16 - P11 = K19 \cdot Q19^K$

$P11 - P10 = K10 \cdot Q10^K$

$P15 - P9 = K9 \cdot Q9^K$

$P15 - P8 = K17 \cdot Q17^K$

$P8 - P7 = K7 \cdot Q7^K$

$P14 - P6 = K6 \cdot Q6^K$

$P14 - P5 = K15 \cdot Q15^K$

$P5 - P4 = K4 \cdot Q4^K$

$P13 - P3 = K3 \cdot Q3^K$

$P13 - P2 = K13 \cdot Q13^K$

$P2 - P1 = K1 \cdot Q1^K$

;NODAL EQUATIONS

$Q13 = Q1 + Q2$

$Q14 = Q3 + Q13$

$Q15 = Q5 + Q4$

$Q16 = Q15 + Q6 + Q14$

$Q17 = Q8 + Q7$

$Q18 = Q17 + Q16 + Q9$

$Q19 = Q11 + Q10$

$Q20 = Q12 + Q19 + Q18$

;PRESSURE EQUATIONS

$P1 = (Q1/KS)^2 + HE$

$P2 = (Q2/KS)^2 + HE$

Rule

$P3 = (Q3/KS)^2 + HE$

$P4 = (Q4/KS)^2 + HE$

$P5 = (Q5/KS)^2 + HE$

$P6 = (Q6/KS)^2 + HE$

$P7 = (Q7/KS)^2 + HE$

$P8 = (Q8/KS)^2 + HE$

$P9 = (Q9/KS)^2 + HE$

$P10 = (Q10/KS)^2 + HE$

$P11 = (Q11/KS)^2 + HE$

$P12 = (Q12/KS)^2 + HE$

Status	Input	Name	Output	Unit	Comment
		K1			PIPE CONSTANT, TYPICAL
	15	L1		FEET	LENGTH OF PIPE 1, TYPICAL
	130	C			HAZEN WILLIAMS COEFFICIENT
	1	D1		INCHES	PIPE 1 ID, TYPICAL
		K3			
	12	L3			
	1	D3			
		K4			
	20	L4			
	1.25	D4			
		K6			
	18	L6			
	1	D6			
		K7			
	15	L7			
	1	D7			
		K9			
	10	L9			
	1	D9			
		K10			
	15	L10			
	1	D10			
		K12			

Variables

Status	Input	Name	Output	Unit	Comment
	5	L12			
	1	D12			
		K13			
	10	L13			
	1	D13			
		K14			
	20	L14			
	1.25	D14			
		K15			
	6	L15			
	1	D15			
		K16			
	20	L16			
	1.5	D16			
		K17			
	15	L17			
	1	D17			
		K18			
	20	L18			
	2	D18			
		K19			
	15	L19			
	1	D19			

Variables

Variables

Status	Input	Name	Output	Unit	Comment
		K20			
	10	L20			
	2	D20			
Guess	1	Q1		GPM	FLOW FROM HEAD 1, TYPICAL
		V1		FT/SEC	FLOW VELOCITY, PIPE 1, TYPICAL
		V3			
Guess	1	Q3			
		V4			
Guess	1	Q4			
		V6			
Guess	1	Q6			
		V7			
Guess	1	Q7			
		V9			
Guess	1	Q9			
		V10			
Guess	1	Q10			
		V12			
Guess	1	Q12			
		V13			
		Q13			
		V14			
		Q14			

8.1 12 head tree.tkw

Variables

Status	Input	Name	Output	Unit	Comment
		V15			
		Q15			
		V16			
		Q16			
		V17			
		Q17			
		V18			
		Q18			
		V19			
		Q19			
		V20			
Guess	1	Q20		GPM	TOTAL FLOW
Guess	1	P17		PSI	TOTAL HEAD REQUIRED
	1.85	P16			
		K			FLOW EXPONENT
Guess	1	P15			
Guess	1	P14			
Guess	1	P13			
Guess	1	P12			
Guess	1	P11			
Guess	1	P10			
Guess	1	P9			
Guess	1	P8			

Variables

Status	Input	Name	Output	Unit	Comment
Guess	1	P7			
Guess	1	P6			
Guess	1	P5			
Guess	1	P4			
Guess	1	P3			
Guess	1	P2			
	10.38	P1		PSI	7 PSI (CODE) + HE (ELEVATION)
Guess	1	Q2			
Guess	1	Q5			
Guess	1	Q8			
Guess	1	Q11			
	5.44	KS			SPRINKLER K-FACTOR
	5.19	HE		PSI	ELEVATION HEAD

8.1 12 head tree.tkw

Variables

Status	Input	Name	Output	Unit	Comment
		K1	.008325963		PIPE CONSTANT, TYPICAL
	15	L1		FEET	LENGTH OF PIPE 1, TYPICAL
	130	C			HAZEN WILLIAMS COEFFICIENT
	1	D1		INCHES	PIPE 1 ID, TYPICAL
		K3	.006660771		
	12	L3			
	1	D3			
		K4	.0037485		
	20	L4			
	1.25	D4			
		K6	.009991156		
	18	L6			
	1	D6			
		K7	.008325963		
	15	L7			
	1	D7			
		K9	.005550642		
	10	L9			
	1	D9			
		K10	.008325963		
	15	L10			
	1	D10			
		K12	.002775321		

8.1 12 head tree.tkw

Status	Input	Name	Output	Unit	Comment
	5	L12			
	1	D12			
		K13	.005550642		
	10	L13			
	1	D13			
		K14	.0037485		
	20	L14			
	1.25	D14			
		K15	.003330385		
	6	L15			
	1	D15			
		K16	.001543836		
	20	L16			
	1.5	D16			
		K17	.008325963		
	15	L17			
	1	D17			
		K18	.000380813		
	20	L18			
	2	D18			
		K19	.008325963		
	15	L19			
	1	D19			

Variables

8.1 12 head tree.tkw

Status	Input	Name	Output	Unit	Comment
		K20	.000190407		
	10	L20			
	2	D20			
		Q1	12.3931749	GPM	FLOW FROM HEAD 1, TYPICAL
		V1	5.06261194	FT/SEC	FLOW VELOCITY, PIPE 1, TYPICAL
		V3	6.03082174		
		Q3	14.7633335		
		V4	4.27438816		
		Q4	16.3494039		
		V6	7.01023827		
		Q6	17.160926		
		V7	7.13768128		
		Q7	17.472904		
		V9	9.06852968		
		Q9	22.1995831		
		V10	7.90711855		
		Q10	19.3564714		
		V12	10.2579713		
		Q12	25.1113127		
		V13	10.5361122		
		Q13	25.7921964		
		V14	10.6028377		
		Q14	40.5555299		

Status	Input	Name	Output	Unit	Comment
		V15	13.5967846		
		Q15	33.2846623		
		V16	16.5217586		
		Q16	91.0011182		
		V17	14.8266141		
		Q17	36.295261		
		V18	15.2672751		
		Q18	149.495962		
		V19	16.415936		
		Q19	40.1858898		
		V20	21.935752		
		Q20	214.793165	GPM	TOTAL FLOW
		P17	31.5027665	PSI	TOTAL HEAD REQUIRED
		P16	27.5770436		
	1.85	K			FLOW EXPONENT
		P15	23.5612046		
		P14	17.0623375		
		P13	13.5243996		
		P12	26.4979188		
		P11	19.8507602		
		P10	17.8506085		
		P9	21.8429752		
		P8	17.1615454		

Variables

8.1 12 head tree.tkw

Status	Input	Name	Output	Unit	Comment
		P7	15.5065		
		P6	15.1413875		
		P5	14.8813852		
		P4	14.22246		
		P3	12.5549713		
		P2	11.256642		
	10.38	P1		PSI	7 PSI (CODE) + HE (ELEVATION)
		Q2	13.3990215		
		Q5	16.9352584		
		Q8	18.822357		
		Q11	20.8294185		
	5.44	KS			SPRINKLER K-FACTOR
	5.19	HE		PSI	ELEVATION HEAD

Variables

8.1 12 head tree.tkw

PROBLEM 8.2 – STANDPIPE SYSTEM

Standpipes are used mainly by the fire department. They can usually be found in stairwells and may have hose stations provided. Standpipes can be wet or dry. Dry standpipes are filled by the fire department from a fire hydrant. The standpipe system is just a divided flow problem and is relatively simple. Minimum pressure at most remote hose station = 100 psi @ 500 gpm. Standpipe systems usually require a fire pump because of these requirements, especially if there is a monitor (fire hose station) on the roof. See the equations in the diagram. Note that pressure required to boost the water up = friction loss + elevation loss + orifice loss at the hose station. Pressure required at the source (fire truck or fire pump) is approximately 196 psi.

STANDPIPES

8.2

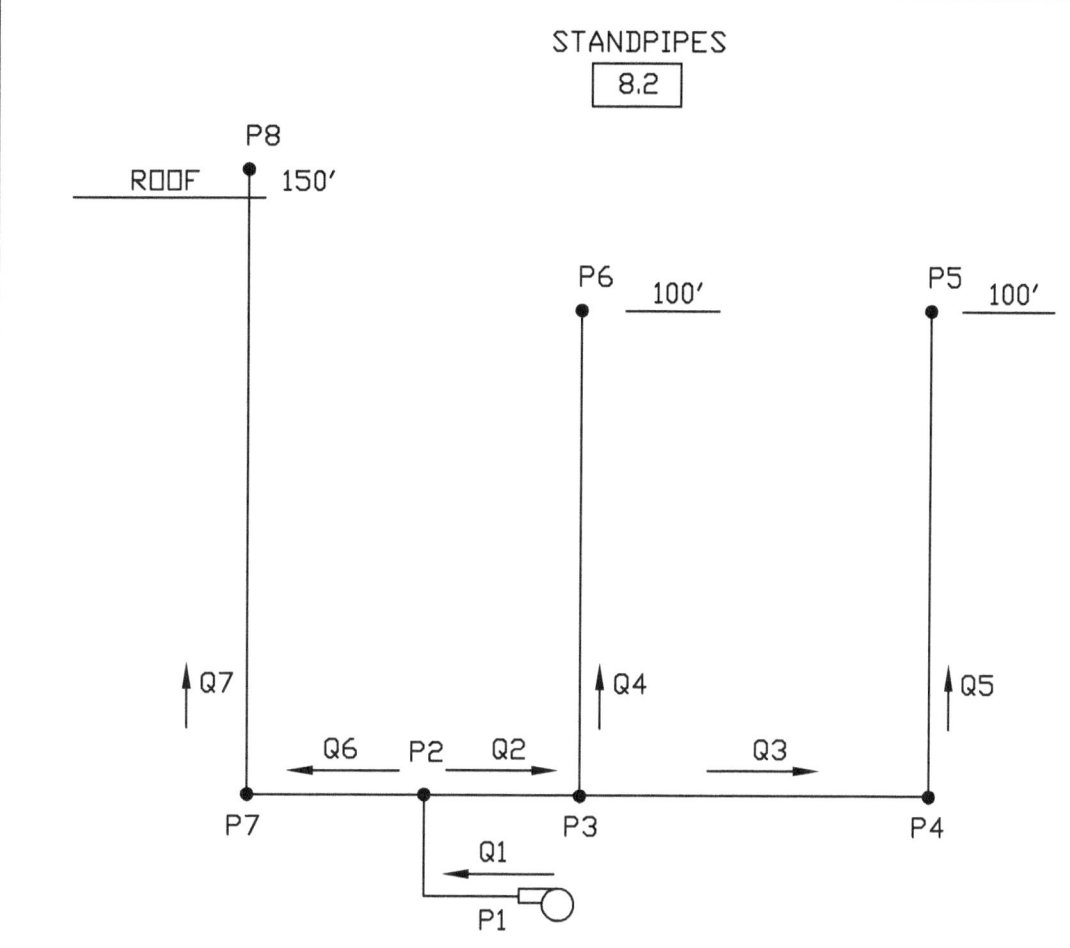

PIPE EQUATIONS:

$P1-P2 = K1 \ast Q1^K$
$P2-P3 = K2 \ast Q2^K$
$P3-P4 = K3 \ast Q3^K$
$P3-P6 = (100/2.31) + K4 \ast Q4^K + (Q4/KV)^2$
$P4-P5 = (100/2.31) + K5 \ast Q5^K + (Q5/KV)^2$
$P2-P7 = K6 \ast Q6^K$
$P7-P8 = (150/2.31) + K7 \ast Q7^K + (Q7/KV)^2$

NODAL EQUATIONS:

$Q1 = Q2 + Q6$
$Q2 = Q4 + Q3$
$Q3 = Q5$
$Q6 = Q7$

8-20

Rule

; TK SOLVER

; 8.2 STANDPIPES 8.02 USING HAZEN WILLIAMS

K1=4.52*L1/((C^1.85)*D1^4.87)

K2=4.52*L2/((C^1.85)*D2^4.87)

K3=4.52*L3/((C^1.85)*D3^4.87)

K4=4.52*L4/((C^1.85)*D4^4.87)

K5=4.52*L5/((C^1.85)*D5^4.87)

K6=4.52*L6/((C^1.85)*D6^4.87)

K7=4.52*L7/((C^1.85)*D7^4.87)

K=1.85

C=130

;FLOW VELOCITIES

V1=0.4085*Q1/(D1^2)

V2=0.4085*Q2/(D2^2)

V3=0.4085*Q3/(D3^2)

V4=0.4085*Q4/(D4^2)

V5=0.4085*Q5/(D5^2)

V6=0.4085*Q6/(D6^2)

V7=0.4085*Q7/(D7^2)

;FLOW EQUATIONS

P1-P2=K1*Q1^K

P2-P3=K2*Q2^K

P3-P4=K3*Q3^K

P3-P6=(100/2.31)+K4*Q4^K

P4-P5=(100/2.31)+K5*Q5^K

P7-P8=(150/2.31)+K7*Q7^K

P2-P7=K6*Q6^K

;NODAL EQUATIONS

Q1=Q2+Q6

Q2=Q4+Q3

Status	Input	Name	Output	Unit	Comment
		K1			
		L1		FEET	PIPE 1 LENGTH, TYPICAL
		C			HAZEN WILLIAMS PIPE COEFFICIENT
		D1		INCHES	PIPE 1 DIA., TYPICAL
		K2			
		L2			
		D2			
		K3			
		L3			
		D3			
		K4			
		L4			
		D4			
		K5			
		L5			
		D5			
		K6			
		L6			
		D6			
		K7			
		L7			
		D7			
		K			

Variables   8.2 STANDPIPES.tkw

Status	Input	Name	Output	Unit	Comment
Guess	1	V1		FT/SEC	FLOW VELOCITY IN PIPE 1
		Q1		GPM	TOTAL FLOW
Guess	1	V2			
		Q2		GPM	
Guess	1	V3			
		Q3		GPM	
Guess	1	V4		FT/SEC	FLOW VELOCITY TO P6
		Q4		GPM	FLOW FROM 1ST STANDPIPE
Guess	1	V5		FT/SEC	FLOW VELOCITY TO P5
		Q5		GPM	FLOW FROM 2ND STANDPIPE
Guess	1	V6			
		Q6		GPM	FLOW FROM MOST REMOTE STANDPIPE
Guess	1	V7		FT/SEC	FLOW VELOCITY TO P8
		Q7		GPM	FLOW FROM MOST REMOTE STANDPIPE
Guess	1	P1		PSI	PRESSURE AT SOURCE
Guess	1	P2			
Guess	1	P3			
Guess	1	P4			
Guess	1	P6			
		KV			
Guess	1	P5			
		P7	1	PSI	
	100	P8		PSI	PRESSURE AT MOST REMOTE STANDPIPE

Variables

8.2 STANDPIPES.tkw

Rule

Q3=Q5

Q6=Q7

P5=(Q5/KV)^2

P6=(Q4/KV)^2

P8=(Q7/KV)^2

;CONSTANTS. NOTE THAT CONSTANTS CAN BE PUT IN RULES SECTION

KV=50

L1= 100

L2=34

L3=67

L4=125

L5=125

L6=175

L7=180

D1=6

D3=4

D4=4

D5=4

D6=4

D7=4

D2=6

Q7=500

Status	Input	Name	Output	Unit	Comment
		K1	9.00996E-6		
		L1	100	FEET	PIPE 1 LENGTH, TYPICAL
		C	130		HAZEN WILLIAMS PIPE COEFFICIENT
		D1	6	INCHES	PIPE 1 DIA., TYPICAL
		K2	3.06331E-6		
		L2	34		
		D2	6		
		K3	4.34872E-5		
		L3	67		
		D3	4		
		K4	8.11293E-5		
		L4	125		
		D4	4		
		K5	8.11328E-5		
		L5	125		
		D5	4		
		K6	.000113592		
		L6	175		
		D6	4		
		K7	.000116838		
		L7	180		
		D7	4		
		K	1.85		

Variables

8.2 STANDPIPES.tkw

Status	Input	Name	Output	Unit	Comment
		V1	18.6151422	FT/SEC	FLOW VELOCITY IN PIPE 1
		Q1	1640.50212	GPM	TOTAL FLOW
		V2	12.9415311	FT/SEC	
		Q2	1140.50212	GPM	
		V3	14.4190986	FT/SEC	
		Q3	564.762736	GPM	
		V4	14.6993463	FT/SEC	FLOW VELOCITY TO P6
		Q4	575.739389	GPM	FLOW FROM 1ST STANDPIPE
		V5	14.4190986	FT/SEC	FLOW VELOCITY TO P5
		Q5	564.762736	GPM	FLOW FROM 2ND STANDPIPE
		V6	12.765625	FT/SEC	
		Q6	500	GPM	FLOW FROM MOST REMOTE STANDPIPE
		V7	12.765625	FT/SEC	FLOW VELOCITY TO P8
		Q7	500	GPM	FLOW FROM MOST REMOTE STANDPIPE
		P1	195.602439	PSI	PRESSURE AT SOURCE
		P2	187.614589		
		P3	186.228411		
		P4	180.866579		
		P6	132.572741		
		KV	50.0018933		
		P5	127.573117		
		P7	176.434542	PSI	
	100	P8		PSI	PRESSURE AT MOST REMOTE STANDPIPE

PROBLEM 8.3 COMBINATION STANDPIPE AND SPRINKLER SYSTEM

This problem combines the sprinkler system (8.1) with the standpipe system (8.2). The sprinkler system will be the same as 8.1 with the addition of a standpipe system. See diagram 8.3. P17 will be the start of the sprinkler system on the 5th floor. The 5th floor elevation is given as 50'. With a 10' ceiling the sprinkler heads will then be at 50' + 10' = 60' elevation. 60' = 25.97 psi. Thus for this problem, HE = 25.97. This will be used in the pressure equations for the sprinkler heads. For instance, for the head at P1, P1 = $(P1/KS)^2$ = 25.97. The pipe flow equations are the same as before, with the addition of pipe flows in the standpipes. The nodal equations are the same as before with the addtion of the standpipe flows.

As per NFPA 14, the most remote standpipe must deliver 500 gpm @ 100 psi. The second most remote standpipe must deliver 250 gpm. Pressure requirements for the second standpipe are not stated.

The objective of this problem is to find the total flow rate and discharge pressure of the fire pump in the basement, assuming that both standpipes plus the 5th floor sprinkler system are operating at the same time. The code says the total flow need not exceed 1250 gpm. The total will exceed this, but we will do it anyway, being conservative.

All standpipes will be 6" diameter. As before, for the standpipe discharge, $Q = k*(P)^{.5}$. Then for 500 gpm and 100 psi, k = 50 for each standpipe.

Elevation heads (HE) are calculated as follows:

P18 (most remote standpipe) will be 4' above the roof elevation = 62' + 4' = 66' = 28.57 psi.

P17 is 1' above the 5th floor ceiling = 50'+10'+1' = 61' = 26.4 psi.

P19 is 1' below ground floor elevation = 2' – 1' = 1' = 0.433 psi.

P21 is the same elevation as P19 = 0.433 psi.

P20 (2nd most remote standpipe) is at the same elevation as P18 = 28.57 psi

P22 (at pump discharge) is 1'-6" above the basement floor = -8' + 1.5' = -6.5' = -2.38 psi.

Pipe L21 (from P17 to P18) = 66'+3' (distance from standpipe to hose outlet) –61' = 8'.

Pipe L22 (from P19 to P17) = 61' – 1' = 60'.

Pipe L23 is from 1st standpipe to 2nd standpipe and is given as 100'.

Pipe L24 is from P21 to the 1st standpipe = 66' + 3' – 1' = 68'.

Pipe L25 (from pump discharge to P21) is given as 6'.

We can now calculate the (5) new pipe constants, K21, K22, K23, K24 and K25.

Starting at the fire pump discharge the new pipe flow equations are calculated taking into account the flow friction PLUS the elevation head differences as this is an open to atmosphere system.

P22-P21 = K25*Q25^K + absolute value of the difference in elevation heads = K25*Q25^K + |0.433 - (2.38)| = K25*Q25^K + 1.947. Note that if we say the difference in elevation = .433-(-2.38), we would in effect, be adding the elevation pressures which is not so. The pump must overcome the NET differences in elevation at each point where the water is lifted.

Next, P21-P20 = K24*Q22^K + (28.57-0.433) = K24*Q22^K + 28.137

Next, P21–P19 = K23*Q21^K + 0 (no change in elevation)

Next, P19-P17 = K24*Q21^K + (26.4-0.433) = K24*Q21^K + 25.967

Next, P17-P18 = K21*500^K + (28.57-26.4) = K21*500^K + 2.17 (Note the 500 gpm flow to P18).

Next the pressure equation:
(Q22/50)^2=P20 (same discharge k factor as for the most remote standpipe)

Next the nodal equations:
Q25=Q22+Q21
Q21=Q20 + 500

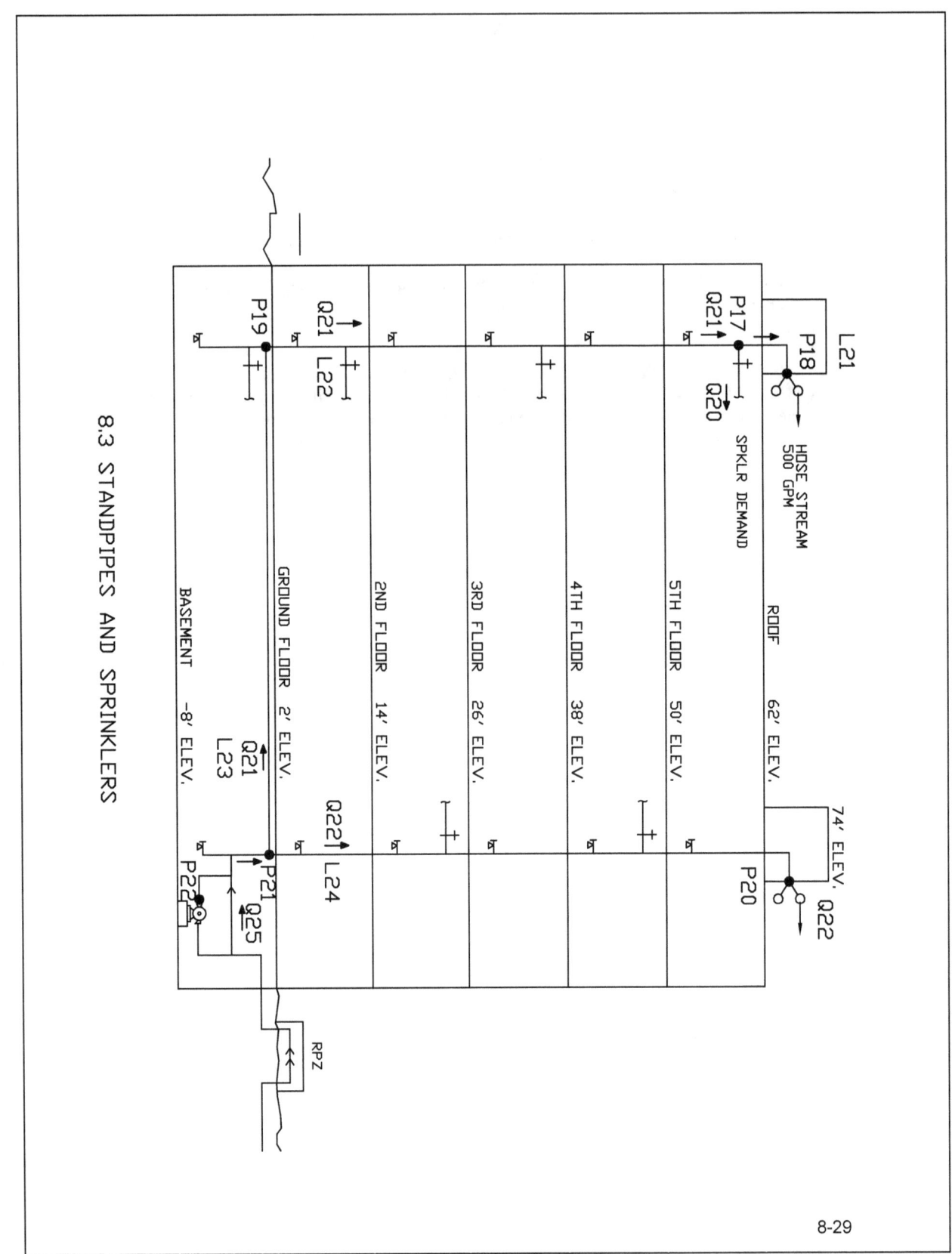

8.3 STANDPIPES AND SPRINKLERS

Rule

;TK SOLVER

;8.3 STANDPIPES AND SPRINKLER SYSTEM

;PIPE CONSTANTS

K1=4.52*L1*C^-1.85*D1^-4.8655
K3=4.52*L3*C^-1.85*D3^-4.8655
K4=4.52*L4*C^-1.85*D4^-4.8655
K6=4.52*L6*C^-1.85*D6^-4.8655
K7=4.52*L7*C^-1.85*D7^-4.8655
K9=4.52*L9*C^-1.85*D9^-4.8655
K10=4.52*L10*C^-1.85*D10^-4.8655
K12=4.52*L12*C^-1.85*D12^-4.8655
K13=4.52*L13*C^-1.85*D13^-4.8655
K14=4.52*L14*C^-1.85*D14^-4.8655
K15=4.52*L15*C^-1.85*D15^-4.8655
K16=4.52*L16*C^-1.85*D16^-4.8655
K17=4.52*L17*C^-1.85*D17^-4.8655
K18=4.52*L18*C^-1.85*D18^-4.8655
K19=4.52*L19*C^-1.85*D19^-4.8655
K20=4.52*L20*C^-1.85*D20^-4.8655
K21=4.52*L21*C^-1.85*D23^-4.8655
K22=4.52*L22*C^-1.85*D22^-4.8655
K23=4.52*L23*C^-1.85*D23^-4.8655
K24=4.52*L24*C^-1.85*D24^-4.8655
K25=4.52*L25*C^-1.85*D25^-4.8655

;COMPUTE VELOCITIES

V1=0.4085*Q1/(D1^2)
V3=0.4085*Q3/(D3^2)
V4=0.4085*Q4/(D4^2)
V6=0.4085*Q6/(D6^2)
V7=0.4085*Q7/(D7^2)
V9=0.4085*Q9/(D9^2)
V10=0.4085*Q10/(D10^2)
V12=0.4085*Q12/(D12^2)

Rule

$V13 = 0.4085 \cdot Q13/(D13^2)$

$V14 = 0.4085 \cdot Q14/(D14^2)$

$V15 = 0.4085 \cdot Q15/(D15^2)$

$V16 = 0.4085 \cdot Q16/(D16^2)$

$V17 = 0.4085 \cdot Q17/(D17^2)$

$V18 = 0.4085 \cdot Q18/(D18^2)$

$V19 = 0.4085 \cdot Q19/(D19^2)$

$V20 = 0.4085 \cdot Q20/(D20^2)$

$V21 = 0.4085 \cdot Q21/(D21^2)$

$V22 = 0.4085 \cdot Q22/(D22^2)$

$V23 = 0.4085 \cdot Q21/(D23^2)$

$V25 = 0.4085 \cdot Q25/(D25^2)$

;PIPE FLOW EQUATIONS

$P17 - P16 = K20 \cdot Q20^K$

$P16 - P15 = K18 \cdot Q18^K$

$P15 - P14 = K16 \cdot Q16^K$

$P14 - P13 = K14 \cdot Q14^K$

$P16 - P12 = K12 \cdot Q12^K$

$P16 - P11 = K19 \cdot Q19^K$

$P16 - P11 = K19 \cdot Q19^K$

$P11 - P10 = K10 \cdot Q10^K$

$P15 - P9 = K9 \cdot Q9^K$

$P15 - P8 = K17 \cdot Q17^K$

$P8 - P7 = K7 \cdot Q7^K$

$P14 - P6 = K6 \cdot Q6^K$

$P14 - P5 = K15 \cdot Q15^K$

$P5 - P4 = K4 \cdot Q4^K$

$P13 - P3 = K3 \cdot Q3^K$

$P13 - P2 = K13 \cdot Q13^K$

$P2 - P1 = K1 \cdot Q1^K$

$P22 - P21 = K25 \cdot Q25^K + 1.947$

$P21 - P20 = K24 \cdot Q22^K + 28.137$

$P21 - P19 = K23 \cdot Q21^K + 0$

$P19 - P17 = K24 \cdot Q21^K + 25.967$

Rule

;NOTE THAT THE FLOW FROM P17 TO P18 IS 500 GPM AS A REQUIREMENT
P17-P18=K21*500^K+2.17

;NODAL EQUATIONS
Q13=Q1+Q2
Q14=Q3+Q13
Q15=Q5+Q4
Q16=Q15+Q6+Q14
Q17=Q8+Q7
Q18=Q17+Q16+Q9
Q19=Q11+Q10
Q20=Q12+Q19+Q18
Q25=Q22+Q21
Q21=Q20+500

;PRESSURE EQUATIONS
P1=(Q1/KS)^2+HE
P2=(Q2/KS)^2+HE
P3=(Q3/KS)^2+HE
P4=(Q4/KS)^2+HE
P5=(Q5/KS)^2+HE
P6=(Q6/KS)^2+HE
P7=(Q7/KS)^2+HE
P8=(Q8/KS)^2+HE
P9=(Q9/KS)^2+HE
P10=(Q10/KS)^2+HE
P11=(Q11/KS)^2+HE
P12=(Q12/KS)^2+HE
(Q22/50)^2=P20

Status	Input	Name	Output	Unit	Comment
		K1	.008325963		
	15	L1		FEET	LENGTH OF PIPE 1, TYPICAL
	130	C			HAZEN WILLIAMS COEFFICIENT
	1	D1		INCHES	PIPE 1 ID, TYPICAL
		K3	.006660771		
	12	L3			
	1	D3			
		K4	.0037485		
	20	L4			
	1.25	D4			
		K6	.009991156		
	18	L6			
	1	D6			
		K7	.008325963		
	15	L7			
	1	D7			
		K9	.005550642		
	10	L9			
	1	D9			
		K10	.008325963		
	15	L10			
	1	D10			
		K12	.002775321		

Variables

8.3 standpipes and sprinklers.tkw

Status	Input	Name	Output	Unit	Comment
	5	L12			
	1	D12			
		K13	.005550642		
	10	L13			
	1	D13			
		K14	.0037485		
	20	L14			
	1.25	D14			
		K15	.003330385		
	6	L15			
	1	D15			
		K16	.001543836		
	20	L16			
	1.5	D16			
		K17	.008325963		
	15	L17			
	1	D17			
		K18	.000380813		
	20	L18			
	2	D18			
		K19	.008325963		
	15	L19			
	1	D19			

Variables

8.3 standpipes and sprinklers.tkw

Status	Input	Name	Output	Unit	Comment
		K20	.000190407		
	10	L20			
	2	D20			
		Q1	22.0377821	GPM	FLOW FROM HEAD 1, TYPICAL
		V1	9.002434	FT/SEC	FLOW VELOCITY, PIPE 1, TYPICAL
		V3	10.5977256		
		Q3	25.9430247		
		V4	7.47950736		
		Q4	28.6088868		
		V6	12.224843		
		Q6	29.9261764		
		V7	12.4747393		
		Q7	30.5379175		
		V9	15.6111372		
		Q9	38.2157581		
		V10	13.7584957		
		Q10	33.6805279		
		V12	17.554874		
		Q12	42.9739877		
		V13	18.6771724		
		Q13	45.7213522		
		V14	18.7359347		
		Q14	71.6643769		

Variables

Status	Input	Name	Output	Unit	Comment
		V15	23.7590604		
		Q15	58.1617146		
		V16	29.0039117		
		Q16	159.752268		
		V17	25.8380843		
		Q17	63.2511243		
		V18	26.6770057		
		Q18	261.21915		
		V19	28.4832165		
		Q19	69.7263561		
		V20	38.1865283		
		Q20	373.919494	GPM	TOTAL SPRINKLER FLOW
		P17	102.241521	PSI	PRESSURE AT SPRINKLER CONNECTION POINT
		P16	91.2938612		
	1.85	K			HAZEN WILLIAMS FLOW EXPONENT
		P15	80.0174461		
		P14	61.6106494		
		P13	51.4675958		
		P12	88.3781555		
		P11	69.8788215		
		P10	64.3058677		
		P9	75.3240002		
		P8	62.135666		

Status	Input	Name	Output	Unit	Comment
		P7	57.4863677		
		P6	56.2364903		
		P5	55.4861118		
		P4	53.6309392		
		P3	48.7167731		
		P2	44.9278104		
		P1	42.3851106	PSI	7 PSI (CODE) IS MINIMUM + HE
		Q2	23.6835701		
		Q5	29.5528278		
		Q8	32.7132068		
		Q11	36.0458281		
	5.44	KS		GPM/(PSI^.5)	SPRINKLER MODIFIED K-FACTOR
	25.974	HE		PSI	ELEVATION HEAD OF SPRINKLERS
		K21	7.26672E-7		
	100	L23			
	6	D23			
		K22	5.45004E-6		
	60	L22			
	6	D22			
		K23	9.0834E-6		
		K24	6.17671E-6		
	68	L24			
	6	D24			

Variables
8.3 standpipes and sprinklers.tkw

Status	Input	Name	Output	Unit	Comment
		K25	5.45004E-7		
	6	L25			
	6	D25			
		P22	134.727507	PSI	PUMP DISCHARGE PRESSURE REQUIRED
		P21	132.428203		
		Q25	1382.99406	GPM	PUMP FLOW RATE
		P19	129.916487		
		Q21	873.919494		
	100	P18		PSI	MINIMUM PRESSURE AT MOST REMOTE STANDPIPE
		P20	103.662707	PSI	CALC PRESSURE AT 2ND MOST REMOTE STANDPIPE
		Q22	509.074569	GPM	FLOW FROM SECOND STANDPIPE
	8	L21		FEET	PIPE LENGTH FROM P17 TO MOST REMOTE STANDPIPE
		V21	9.9165587		
	6	D21			
		V22	5.77658226		
		V23	9.9165587		
		V25	15.693141		

Variables

8.3 standpipes and sprinklers.tkw

CHAPTER 9 – PUMPS IN NETWORKS

DISCUSSION
FIGURE 9.1 - TACO PUMP CATALOG EXCERPT

FIGURE 9.2 - TACO PUMP CURVE FIT

FIGURE 9.3 - SINGLE PUMP DIAGRAM

FIGURE 9.4 - SINGLE PUMP RULES

FIGURE 9.5 - SINGLE PUMP SOLUTION

FIGURE 9.6 - 2 PUMPS IN SERIES DIAGRAM

FIGURE 9.7 – 2 PUMPS IN SERIES RULES

FIGURE 9.8 – 2 PUMPS IN SERIES SOLUTION

FIGURE 9.9 – 2 PUMPS IN PARALLEL DIAGRAM

FIGURE 9.10 – 2 PUMPS IN PARALLEL RULES

FIGURE 9.11 – 2 PUMPS IN PARALLEL SOLUTION

FIGURE 9.12 – GEOTHERMAL WELL SYSTEM DIAGRAM

FIGURE 9.13 – GEOTHERMAL WELL SYSTEM RULES

FIGURE 9.14 – GEOTHERMAL WELL SYSTEM SOLUTION

FIGURE 9.15 – TACO 133 PUMP CATALOG EXCERPT

FIGURE 9.16 – TACO 133 PUMP CURVE

FIGURE 9.17 – PUMPS, CONDENSERS AND COOLING TOWERS (PCCT) DIAGRAM

FIGURE 9.18 – PCCT RULES

FIGURE 9.19 – PCCT SOLUTION

FIGURE 9.20 – LIBERTY FL60 PUMP

FIGURE 9.21 – LIBERTY FL60 PUMP CURVE

FIGURE 9.22 – CONDENSATE PUMPS IN PARALLEL DIAGRAM

FIGURE 9.23 – CONDENSATE PUMPS IN PARALLEL RULES & SOLUTION

FIGURE 9.24 – LITTLE GIANT PERISTALTIC POSITIVE DISPLACEMENT PUMP

FIGURE 9.25 – POSITIVE DISPLACEMENT CONDENSATE PARALLEL PUMPS DIAGRAM

FIGURE 9.26 – POSITIVE DISPLACEMENT CONDENSATE PARALLEL PUMPS GIVEN PRESSURES RULES & SOLUTION

FIGURE 9.27 – POSITIVE DISPLACEMENT CONDENSATE PARALLEL PUMPS GIVEN FLOWS RULES & SOLUTION

FIGURE 9.28 – THE STRANGE CASE OF GRAVITY LINE CONNECTED TO A PUMPED LINE DIAGRAM

FIGURE 9.29 – GRAVITY LINE TO PUMPED LINE RULES & SOLUTION

FINAL COMMENTS

DISCUSSION

What is a pump? It may sound trivial, but what is it really but a reservoir with a variable head? With that in mind, it follows that pumps can be modeled as part of a piping network. We have already worked several problems given the inlet pressure. The key is to have a mathematical model of a pump's performance as close as practical to the actual pump curve. This is done by finding an equation that provides the discharge pressure given the pump flow rate. To find this equation, we perform a curve fit, otherwise known as regression analysis. Note that positive displacement pumps do NOT have a variable head. This will be addressed later.

The 1st pump chosen is a TACO-133 circulation pump. See pump curve, fig. 9.1. The data was taken from the curve and shows up in the TK Solver curve fit, fig. 9.2. The accuracy = 0.9994 or, 99.94%, plenty accurate for our use. The equation is $p = b1 + b2*q + b3*q^2$ where p = ft. of head, and q = gpm. Note that p is discharge head. Anything immediately downstream creates pressure drop.

This pump will be used in the first problem, Figure 9.3. The pressure head on the pump will be calculated from the pump discharge to the suction. There is a gate valve and check valve immediately downstream of the pump, plus piping and two cooling coils. Note that the discharge pressure = suction pressure + discharge pressure calculated from the pump curve equation. Note that the gate valves and check valves are modeled by adding the Kg and Kg factors to the equations. Pressure drop for the valves = $Kg*V^2/(2*g) + Kc*V^2/(2*g)$.

Figure 9.4- Two pumps in series in network:

This is handled in a similar way to 1 pump in the network. The suction pressure of the 2nd pump is the discharge pressure of the 1st pump – minor losses of a gate valve and check valve. The flow rate is the same, obviously, for there is nowhere else for the water to go. Adding a second pump is a way to boost pressure while keeping the flow rates the same if additional pressure is required.

Fig. 9.5 – Two pumps in parallel in network:

This is handled differently from the previous problem. The flows are different for each pump. The flow will divide according to the pressure drop in the circuit for each pump. The flow rates are slightly different because of the longer circuit length for pump 1.

Fig. 9.6 – GEOTHERMAL SYSTEM with 1 pump.. This is an interesting circuit with water source heat pumps in parallel and geothermal wells in parallel also. Each well is 300' deep, with 2 tees and 4 elbows to produce an equivalent length of 626'. The wells are 15' on center. The circuit is modeled with HDPE piping with a relative roughness of 0.00015. The water enters the wells at 95 Deg.F and leaves at 85 Deg.F for an average temperature of 90 Deg.F. This is used for the water viscosity and density. The total tonnage is known, and so is the total flow, 28 gpm, based on 3.1 gpm/ton. The required head across the pump is calculated and so the pump can be selected.

Fig. 9.7 – PUMPS, CONDENSERS, COOLING TOWERS (PCCT). Up to now, all the pumping systems have been closed. This one is open to atmosphere at the cooling towers. Thus we must take account of the elevation difference between the condensers and the cooling towers. The pump curve is modeled after the TACO TA Series Model 2530, 1160 RPM. Note the different flow rates for each pump.

Now we consider condensate pumps, a very common pump application in HVAC design work. These pumps will be in parallel, so that all of them pump into a common header. See Figures 9.20, 21 and 22. Note elevation of P8, the termination of the line. This would be similar to a condensate line ending 1' above a floor drain or sink that has a funnel attached. A centrifugal sump pump (Liberty FL60) was modeled.

A similar problem was done using Peristaltic pumps, which are positive discharge. See Figures 9.25 through 9.27 for solutions holding pressure constant and then flow constant. The pumps can easily handle whatever condensate the A/C units can produce.

Now we come to the strange case of a gravity line connected to a pumped line. This is not fiction. It actually occurred at a client's facility. A pumped line was in place, piping condensate waste to a disposal point. Later, an A/C unit was added downstream of the condensate pump. Apparently, whoever designed the system thought it was okay to join the gravity condensate line from the new A/C unit to the pumped line. **Wrong.** This would only work if the gravity line had enough head to at least equal the head developed by the pump or if the pump was not running. Actually, it was just about 5' above the pumped line. The solution shows the flow Q3 to be negative, meaning that it is opposite the assumed direction.

The pump would blast condensate mostly up into the new A/C unit's condensate pan and the rest to the disposal point. A waterfall would (and did) occur.

The solution is of course to route the condensate line from the new A/C unit to the suction connection of the condensate pump.

Submittal Data Information
121-138 Series Circulators

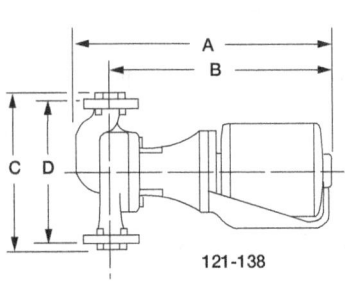

121-138

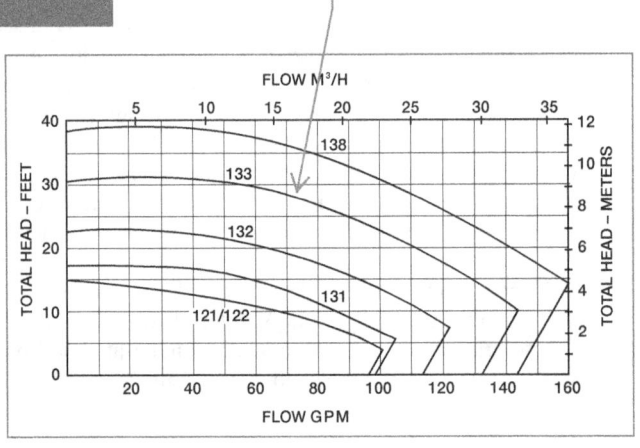

Use this curve

Specifications (121-138 Series)

MODEL NUMBER[1]	FLANGE SIZE	MOTOR[2] 60C/AC 1 PH	RPM	DIMENSIONS								SHIPPING WEIGHT	
				A		B		C		D			
				IN.	MM	IN.	MM	IN.	MM	IN.	MM	LBS.	KG
121	2½"	¼ HP, 115 V	1725	18⅛	460.4	15⅞	403.2	14¼	362.0	11⅛	282.6	72	32.7
122	3"	¼ HP, 115 V	1725	18⅛	460.4	15⅞	403.2	13⅝	346.1	11⅛	282.6	72	32.7
131	3"	⅓ HP, 115 V	1725	19¼	489.0	15¾	400.1	16	406.4	13⅝	346.1	95	43.1
132	3"	½ HP, 115/230 V or 230/460/60/3 or 200/60/3	1725	21½	546.1	18	457.2	16	406.4	13⅝	346.1	108	49.0
133	3"	¾ HP, 115/230 V or 230/460/60/3 or 200/60/3	1725	22⅛	562.0	18⅝	473.1	16	406.4	13⅝	346.1	113	51.3
138	3"	1 HP, 115/230 V or 230/460/60/3 or 200/60/3	1725	22⅝	574.0	19⅛	485.8	16	406.4	13⅝	346.1	118	53.5

(1) When specifying all bronze construction, add letter "B" after model number (i.e. 132B).
When specifying bronze fitted construction, add letter "C" after model number (i.e. 132C).
(2) Motors are available with other electrical characteristics – consult your Taco representative.

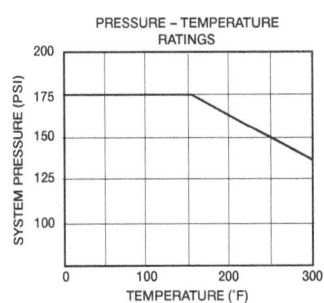

FIGURE 9.1

Taco Inc., 1160 Cranston Street, Cranston, RI 02920 / (401) 942-8000 / Fax (401) 942-2360
Taco (Canada) Ltd., 8450 Lawson Road, Unit #3, Milton, Ontario L9T 0J8 / (905) 564-9422 / Fax (905) 564-9436
www.taco-hvac.com

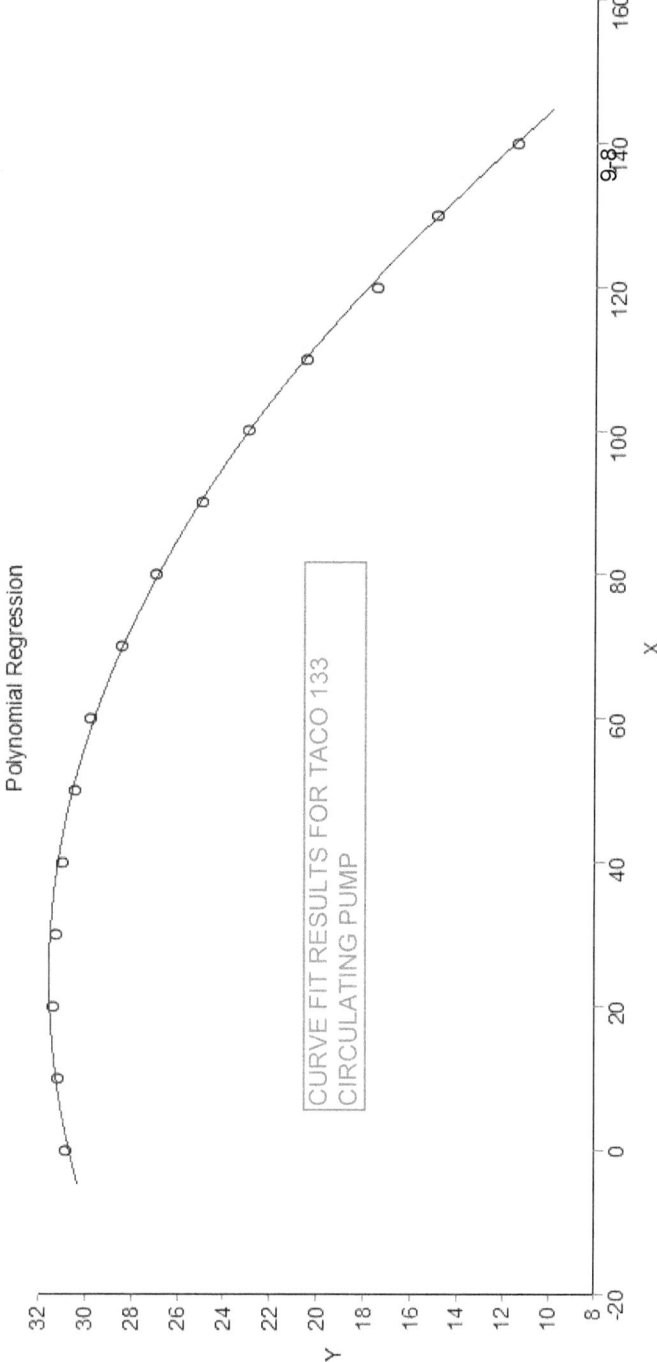

9.3 ONE PUMP IN NETWORK

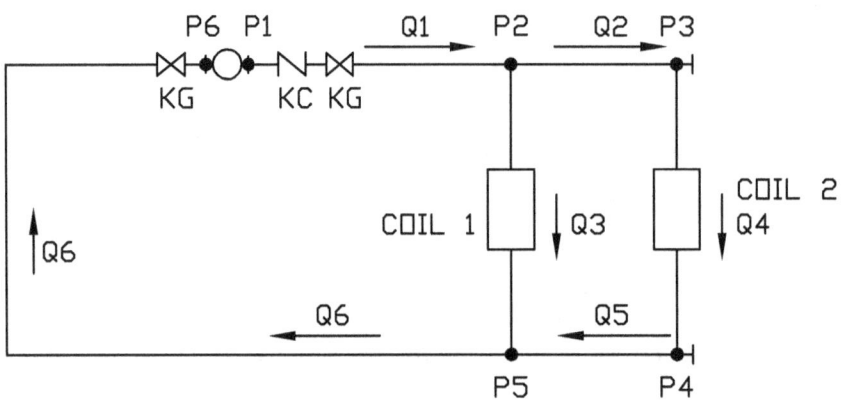

FLOW EQUATIONS:

$P1-P2 = K1*Q1\wedge K+(KC+KG)*V1^2/(2*g)$
$P2-P3 = K2*Q2\wedge K$
$P2-P5 = K3*Q3\wedge K+(Q3/KC1)^2$
$P3-P4 = K4*Q4\wedge K+(Q4/KC2)^2$
$P4-P5 = K5*Q5^2$
$P5-P6 = K6*Q6\wedge K+KG*V6^2/(2*g)$

NODAL EQUATIONS:

Q1=Q2+Q3
Q2=Q4
Q4=Q5
Q3+Q5=Q6
Q6=Q1

PUMP CURVE:

$P1 = b1+b2*Q1+b3*Q1^2$
b1=30.6848629
b2=0.071243051
b3=-0.00148796

GIVEN:

K COIL 1 = 30/(10^.5) = 9.48
K COIL 2 = 25/(15^.5) = 6.45

9-9

Rule

FIGURE 9.3
1 PUMP IN NETWORK
RULES

;>>>/* TK SOLVER */
;>>>/* NAME OF PROBLEM */
;>>>/* 07/18/14 */
;>>>/* PROBLEM USING HAALAND EQUATION FOR FRICTION FACTOR */

;>>>/* CURVE FIT FOR WATER VISCOSITY */
v=a1+a2*T+a3*T^2+a4*T^3+a5*T^4+a6*T^5
a1=5.5728022*10^-4
a2=0.22430157
a3=-9.1107004*10^-3
a4=1.5542381*10^-4
a5=-1.2390396*10^-6
a6=3.7904162*10^-9

;>>>/* CURVE FIT FOR WATER DENSITY */
rho=b1+b2*T+b3*T^2+b4*T^3
b1=62.2614555
b2=9.339695810^-3
b3=-1.37041*10^-4
b4=1.83106*10^-7

;>>>/* CALCULATE REYNOLDS NUMBERS */
R1=50.6*Q1*rho/(D1*v)
R2=50.6*Q2*rho/(D2*v)
R3=50.6*Q3*rho/(D3*v)
R4=50.6*Q4*rho/(D4*v)
R5=50.6*Q5*rho/(D5*v)
R6=50.6*Q6*rho/(D6*v)

;>>>/* CALCULATE f BY SHACHAM EQUATION */
f1=(-2*log((e*12/D1)/3.7-(5.02/R1)*log((e*12/D1)/3.7+14.5/R1)))^-2
f2=(-2*log((e*12/D2)/3.7-(5.02/R2)*log((e*12/D2)/3.7+14.5/R2)))^-2
f3=(-2*log((e*12/D3)/3.7-(5.02/R3)*log((e*12/D3)/3.7+14.5/R3)))^-2
f4=(-2*log((e*12/D4)/3.7-(5.02/R4)*log((e*12/D4)/3.7+14.5/R4)))^-2
f5=(-2*log((e*12/D5)/3.7-(5.02/R5)*log((e*12/D5)/3.7+14.5/R5)))^-2
f6=(-2*log((e*12/D6)/3.7-(5.02/R6)*log((e*12/D6)/3.7+14.5/R6)))^-2

;>>>/* CALCULATE VELOCITY IN EACH PIPE */
V1=0.4085*Q1/(D1^2)
V2=0.4085*Q2/(D2^2)
V3=0.4085*Q3/(D3^2)
V4=0.4085*Q4/(D4^2)
V5=0.4085*Q5/(D5^2)
V6=0.4085*Q6/(D6^2)

;>>>/* CALCULATE K FACTOR FOR EACH PIPE */
K1=0.03108*f1*L1/(D1^5)
K2=0.03108*f2*L2/(D2^5)
K3=0.03108*f3*L3/(D3^5)

Rule
K4=0.03108*f4*L4/(D4^5)
K5=0.03108*f5*L5/(D5^5)
K6=0.03108*f6*L6/(D6^5)

;>>>/* CURVE FIT FOR TACO 133 PUMP */
P1=C1+C2*Q1+C3*Q1^2+P6
C1=30.6848629
C2=0.071243051
C3=-0.00148796

;>>>/* PIPE EQUATIONS */
P1-P2=K1*Q1^K+(KC+KG)*V1^2/(2*G)
P2-P3=K2*Q2^K
P2-P5=K3*Q3^K+(Q3/KC1)^2
P3-P4=K4*Q4^K+(Q4/KC2)^2
P4-P5=K5*Q5^K
P5-P6=K6*Q6^K+KG*V6^2/(2*G)

;>>>/* NODAL EQUATIONS */
Q1=Q2+Q3
Q2=Q4
Q4=Q5
Q3+Q5=Q6
Q6=Q1

;>>>/* INPUT CONSTANTS */
e=0.00005
T=160
K=2
G=32.17
KC1=9.48
KC2=6.45
KG=0.15
KC=2
L1=186
D1=1.5
L2=35
D2=1.5
L3=15
D3=1.5
L4=23
D4=1.5
L5=45
D5=1.5
L6=100
D6=1.5
P6=5

FIG. 9.4 1 PUMP IN NETWORK SOLUTION

Status	Input	Name	Output	Unit	Comment
		v	24.7077574		WATER VISCOSITY AT TEMP T
		a1	.000557280		
		a2	.22430157		
	160	T		DEG.F	WATER TEMPERATURE
		a3	-.0091107		
		a4	.000155423		
		a5	-1.23904E-6		
		a6	3.790416E-9		
		rho	59.6995991	LB/CU.FT.	WATER DENSITY AT TEMP. T
		b1	62.2614555		
		b2	.001227444		
		b3	-.000137041		
		b4	1.83106E-7		
		R1	1770.27222		
		Q1	21.7191443	GPM	PUMP DISCHARGE FLOW
		D1	1.5	INCHES	PIPE 1 I.D., TYPICAL
		R2	605.812986		PIPE 1 REYNOLDS NUMBER, TYPICAL
		Q2	7.43260812		
		D2	1.5		
		R3	1164.45923		
		Q3	14.2865362		
		D3	1.5		
		R4	605.812986		
		Q4	7.43260812		
		D4	1.5		
		R5	605.812986		
		Q5	7.43260812		
		D5	1.5		
		R6	1770.27222		
		Q6	21.7191443		
		D6	1.5		
		f1	.050672307		PIPE 1 FRICTION FACTOR, TYPICAL
		e	.00005		PIPE ROUGHNESS FACTOR
		f2	.071576168		
		f3	.057743312		
		f4	.071576168		
		f5	.071576168		

Variables

9.03 1 pump in network SHACHAM.tkw

Status	Input	Name	Output	Unit	Comment
		f6	.050672307		
		V1	3.94323131	FT./SEC.	PIPE 1 FLOW VELOCITY, TYPICAL
		V2	1.34943129		
		V3	2.59380001		
		V4	1.34943129		
		V5	1.34943129		
		V6	3.94323131		
		K1	.038575213		
		L1	186	FEET	PIPE 1 LENGTH, TYPICAL
		K2	.010253242		
		L2	35		
		K3	.003545011		
		L3	15		
		K4	.006737844		
		L4	23		
		K5	.013182739		
		L5	45		
		K6	.020739362		
		L6	100		
		P1	36.5303056	FT.W.G.	PUMP DISCHARGE PRESSURE
		C1	30.6848629		FACTOR FOR PUMP CURVE FIT
		C2	.071243051		
		C3	-.00148796		
		P2	17.8140801	FT.W.G.	PRESSURE AT POINT 3, TYPICAL
		K	2		EXPONENT IN DARCY-WEISBACH DP EQUATION
		KC	2		K FACTOR FOR CHECK VALVE
		KG	.15		K FACTOR FOR GATE VALVE
		G	32.17	FT./SEC^2	ACCELLERATION OF GRAVITY
		P3	17.2476696		
		P5	14.8193895		
		KC1	9.48		
		P4	15.5476317		
		KC2	6.45		
		P6	5	FT. W.G.	PUMP SUCTION PRESSURE, GIVEN

Variables

9.03 1 pump in network SHACHAM.tkw

9.4 TWO PUMPS IN SERIES IN NETWORK

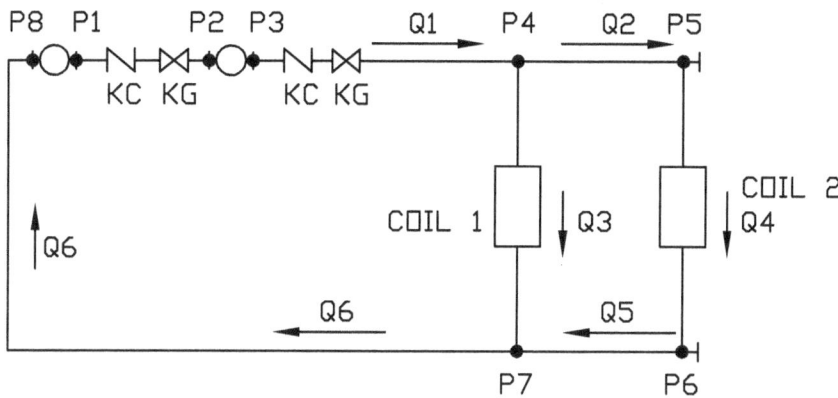

FLOW EQUATIONS:

P1-P2=K1*Q1^K+(KC+KG)*V1^2/(2*g)
P3-P4=K1*Q1^K
P4-P5=K2*Q2^K
P4-P7=K3*Q3^K+(Q3/KC1)^2
P5-P6=K4*Q4^K+(Q4/KC2)^2
P6-P7=K5*Q5^K
P7-P8=K6*Q6^K

NODAL EQUATIONS:

Q1=Q2+Q3
Q2=Q4
Q4=Q5
Q3+Q5=Q6
Q6=Q1

PUMP CURVE:

P1=b1+b2*Q1+b3*Q1^2+P8
b1=30.6848629
b2=0.071243051
b3=-0.00148796

PUMP CURVE:

P3=P2+b1+b2*Q1+b3*Q1^2
b1=30.6848629
b2=0.071243051
b3=-0.00148796

Rule
;>>>/* TK SOLVER */
;>>>/* NAME OF PROBLEM */
;>>>/* 07/18/14 */
;>>>/* PROBLEM USING SHACHAM EQUATION FOR FRICTION FACTOR */

;>>>/* CURVE FIT FOR WATER VISCOSITY */
v=a1+a2*T+a3*T^2+a4*T^3+a5*T^4
a1=3.36899368
a2=-0.059873165
a3=0.000515554
a4=-2.1467*10^-6
a5=3.430662*10^-9

;>>>/* CURVE FIT FOR WATER DENSITY */
rho=b1+b2*T+b3*T^2+b4*T^3
b1=62.2614555
b2=9.339695810^-3
b3=-1.37041*10^-4
b4=1.83106*10^-7

;>>>/* CALCULATE REYNOLDS NUMBERS */
R1=50.6*Q1*rho/(D1*v)
R2=50.6*Q2*rho/(D2*v)
R3=50.6*Q3*rho/(D3*v)
R4=50.6*Q4*rho/(D4*v)
R5=50.6*Q5*rho/(D5*v)
R6=50.6*Q6*rho/(D6*v)
R7=50.6*Q7*rho/(D7*v)
R8=50.6*Q8*rho/(D8*v)
R9=50.6*Q9*rho/(D9*v)
R10=50.6*Q10*rho/(D10*v)
R11=50.6*Q11*rho/(D11*v)
R12=50.6*Q12*rho/(D12*v)
R13=50.6*Q13*rho/(D13*v)
R14=50.6*Q14*rho/(D14*v)
R15=50.6*Q15*rho/(D15*v)

;>>>/* CALCULATE f BY SHACHAM EQUATION */
f1=(-2*log((e*12/D1)/3.7-(5.02/R1)*log((e*12/D1)/3.7+14.5/R1)))^-2
f2=(-2*log((e*12/D2)/3.7-(5.02/R2)*log((e*12/D2)/3.7+14.5/R2)))^-2
f3=(-2*log((e*12/D3)/3.7-(5.02/R3)*log((e*12/D3)/3.7+14.5/R3)))^-2
f4=(-2*log((e*12/D4)/3.7-(5.02/R4)*log((e*12/D4)/3.7+14.5/R4)))^-2
f5=(-2*log((e*12/D5)/3.7-(5.02/R5)*log((e*12/D5)/3.7+14.5/R5)))^-2
f6=(-2*log((e*12/D6)/3.7-(5.02/R6)*log((e*12/D6)/3.7+14.5/R6)))^-2
f7=(-2*log((e*12/D7)/3.7-(5.02/R7)*log((e*12/D7)/3.7+14.5/R7)))^-2
f8=(-2*log((e*12/D8)/3.7-(5.02/R8)*log((e*12/D8)/3.7+14.5/R8)))^-2
f9=(-2*log((e*12/D9)/3.7-(5.02/R9)*log((e*12/D9)/3.7+14.5/R9)))^-2
f10=(-2*log((e*12/D10)/3.7-(5.02/R10)*log((e*12/D10)/3.7+14.5/R10)))^-2
f11=(-2*log((e*12/D11)/3.7-(5.02/R11)*log((e*12/D11)/3.7+14.5/R11)))^-2
f12=(-2*log((e*12/D12)/3.7-(5.02/R12)*log((e*12/D12)/3.7+14.5/R12)))^-2

Rule
f13=(-2*log((e*12/D13)/3.7-(5.02/R13)*log((e*12/D13)/3.7+14.5/R13)))^-2
f14=(-2*log((e*12/D14)/3.7-(5.02/R14)*log((e*12/D14)/3.7+14.5/R14)))^-2
f15=(-2*log((e*12/D15)/3.7-(5.02/R15)*log((e*12/D15)/3.7+14.5/R15)))^-2

;>>>/* CALCULATE VELOCITY IN EACH PIPE */
V1=0.4085*Q1/(D1^2)
V2=0.4085*Q2/(D2^2)
V3=0.4085*Q3/(D3^2)
V4=0.4085*Q4/(D4^2)
V5=0.4085*Q5/(D5^2)
V6=0.4085*Q6/(D6^2)
V7=0.4085*Q7/(D7^2)
V8=0.4085*Q8/(D8^2)
V9=0.4085*Q9/(D9^2)
V10=0.4085*Q10/(D10^2)
V11=0.4085*Q11/(D11^2)
V12=0.4085*Q12/(D12^2)
V13=0.4085*Q13/(D13^2)
V14=0.4085*Q14/(D14^2)
V15=0.4085*Q15/(D15^2)

;>>>/* CALCULATE K FACTOR FOR EACH PIPE */
K1=0.03108*f1*L1/(D1^5)
K2=0.03108*f2*L2/(D2^5)
K3=0.03108*f3*L3/(D3^5)
K4=0.03108*f4*L4/(D4^5)
K5=0.03108*f5*L5/(D5^5)
K6=0.03108*f6*L6/(D6^5)
K7=0.03108*f7*L7/(D7^5)
K8=0.03108*f8*L8/(D8^5)
K9=0.03108*f9*L9/(D9^5)
K10=0.03108*f10*L10/(D10^5)
K11=0.03108*f11*L11/(D11^5)
K12=0.03108*f12*L12/(D12^5)
K13=0.03108*f13*L13/(D13^5)
K14=0.03108*f14*L14/(D14^5)
K15=0.03108*f15*L15/(D15^5)

;>>>/* CURVE FIT FOR TACO 133 PUMP */
P1=P13+C1+C2*Q1+C3*Q1^2
C1=30.6848629
C2=0.071243051
C3=-0.00148796

P3=P14+C1+C2*Q2+C3*Q2^2

;>>>/* PIPE EQUATIONS */
P1-P2=K1*Q1^K+(KC+KG)*V1^2/(2*G)

Rule
P3-P4=K2*Q2^K+(KC+KG)*V2^2/(2*G)
P2-P4=K3*Q3^K
P4-P5=K4*Q4^K
P5-P6=K5*Q5^K
P6-P7=K6*Q6^K
P5-P10=K7*Q7^K+(Q7/KC1)^2
P6-P9=K8*Q8^K+(Q8/KC2)^2
P7-P8=K9*Q9^K+(Q9/KC3)^2
P8-P9=K10*Q10^K
P9-P10=K11*Q11^K
P10-P11=K12*Q12^K
P11-P12=K13*Q13^K
P11-P14=K15*Q15^K+(KG*V15^2/(2*G))
P12-P13=K14*Q14^K+KC*V14^2/(2*G)

;>>>/* NODAL EQUATIONS */
Q1=Q3
Q3+Q2=Q4
Q4=Q7+Q5
Q5=Q8+Q6
Q6=Q9
Q9=Q10
Q10+Q8=Q11
Q7+Q11=Q12
Q12=Q15+Q13
Q15=Q2
Q13=Q14
Q14=Q1

;>>>/* INPUT CONSTANTS */
e=0.00005
T=50
K=2
G=32.17
KC1=16
KC2=16
KC3=10
KG=0.15
KC=2
L1=15
L2=15
L3=20
L4=35
L5=38
L6=46
L7=55
L8=55
L9=45
L10=46
L11=38

Rule
L12=35
L13=20
L14=10
L15=10
D1=3
D2=3
D3=3
D4=3
D5=3
D6=2
D7=2
D8=2
D9=2
D10=2
D11=3
D12=3
D13=3
D14=3
D15=3
P13=5

P1-P13=PD1
P3-P14=PD2

FIGURE 9.7
2 PUMPS IN SERIES SOLUTION

Status	Input	Name	Output	Unit	Comment
		v	24.7077574		WATER VISCOSITY AT TEMP T
		a1	.000557280		
		a2	.22430157		
	160	T		DEG.F	WATER TEMPERATURE
		a3	-.0091107		
		a4	.000155423		
		a5	-1.23904E-6		
		a6	3.790416E-9		
		rho	59.6995991	LB/CU.FT.	WATER DENSITY AT TEMP. T
		b1	62.2614555		
		b2	.001227444		
		b3	-.000137041		
		b4	1.83106E-7		
		R1	3523.42089		
		Q1	57.6376039	GPM	PUMP DISCHARGE FLOW
		D1	2	INCHES	PIPE 1 I.D., TYPICAL
		R2	1302.52855		PIPE 1 REYNOLDS NUMBER, TYPICAL
		Q2	21.3073110		
		D2	2		
		R3	2961.18979		
		Q3	36.3302929		
		D3	1.5		
		R4	1736.70473		
		Q4	21.3073110		
		D4	1.5		
		R5	1736.70473		
		Q5	21.3073110		
		D5	1.5		
		R6	3523.42089		
		Q6	57.6376039		
		D6	2		
		f1	.041313588		PIPE 1 FRICTION FACTOR, TYPICAL
		e	.00005		PIPE ROUGHNESS FACTOR
		f2	.055659973		
		f3	.043509878		
		f4	.050969270		
		f5	.050969270		

Status	Input	Name	Output	Unit	Comment
		f6	.041313588		
		V1	5.886240306	FT./SEC.	PIPE 1 FLOW VELOCITY, TYPICAL
		V2	2.176009133		
		V3	6.595966652		
		V4	3.868460686		
		V5	3.868460686		
		V6	5.886240306		
		K1	.007463403		
		L1	186	FEET	PIPE 1 LENGTH, TYPICAL
		K2	.001892091		
		L2	35		
		K3	.002671184		
		L3	15		
		K4	.004798008		
		L4	23		
		K5	.009387407		
		L5	45		
		K6	.004012582		
		L6	100		
		P1	57.44005699	FT.W.G.	PUMP 1 DISCHARGE PRESSURE
		C1	30.6848629		FACTOR FOR PUMP CURVE FIT
		C2	.071243051		
		C3	-.00148796		
		P2	31.48822798	FT.W.G.	PUMP 2 SUCTION PRESSURE
		K	2		EXPONENT IN DARCY-WEISBACH DP EQUATION
		KC	2		K FACTOR FOR CHECK VALVE
		KG	.15		K FACTOR FOR GATE VALVE
		G	32.17	FT./SEC^2	ACCELLERATION OF GRAVITY
		P3	61.33625486	FT. W. G.	PUMP 2 DISCHARGE PRESSURE
		P5	35.68320546		
		KC1	9.48		
		P4	36.5422190		
		KC2	6.45		
		P6	22.59203011	FT. W.G.	
5		P8		FT. W.G.	PUMP 1 SUCTION PRESSURE, GIVEN
		P7	18.3301268		

9.05 2 PUMPS IN PARALLEL IN NETWORK

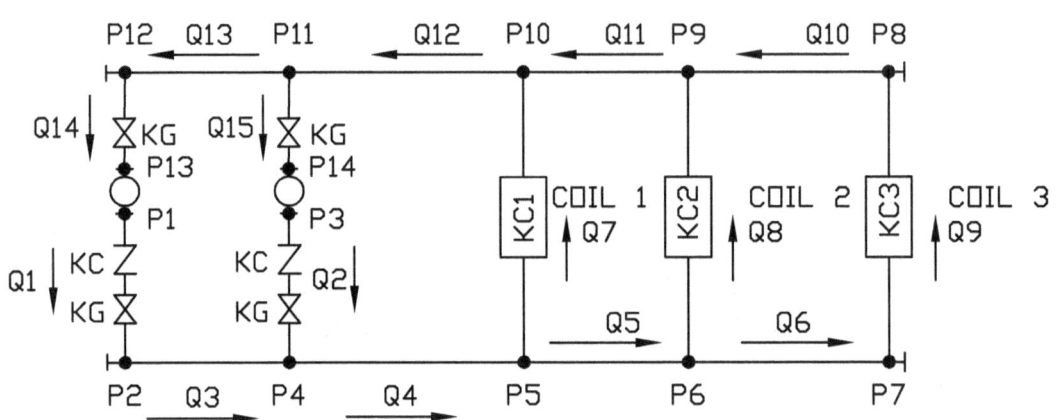

FLOW EQUATIONS:

$P1 = P13 + C1 + C2*Q1 + C3*Q1^2$
$P1 - P2 = K1*Q1^K + (KC+KG)*V1^2/(2*g)$
$P3 = P14 + C1 + C2*Q2 + C3*Q2^2$
$P3 - P4 = K2*Q2^K + (KC+KG)*V2^2/(2*G)$
$P2 - P4 = K3*Q3^K$
$P4 - P5 = K4*Q4^K$
$P5 - P6 = K5*Q5^K$
$P6 - P7 = K6*Q6^K$
$P5 - P10 = K7*Q7^K + (Q7/KC1)^2$
$P6 - P9 = K8*Q8^K + (Q8/KC2)^2$
$P7 - P8 = K9*Q9^K + (Q9/KC3)^2$
$P8 - P9 = K10*Q10^K$
$P9 - P10 = K11*Q11^K$
$P10 - P11 = K12*Q12^K$
$P11 - P12 = K13*Q13^K$
$P11 - P14 = K15*Q15^K + KG*V15^2/(2*G)$
$P12 - P13 = K14*Q14^K + KG*V14^2/(2*G)$

NODAL EQUATIONS:

$Q1 = Q3$
$Q3 + Q2 = Q4$
$Q4 = Q7 + Q5$
$Q5 = Q8 + Q6$
$Q6 = Q9$
$Q9 = Q10$
$Q10 + Q8 = Q11$
$Q7 + Q11 = Q12$
$Q12 = Q15 + Q13$
$Q15 = Q2$
$Q13 = Q14$
$Q14 = Q1$

PUMP CURVE 1:

$P1 = P13 + c1 + c2*Q1 + c3*Q1^2$
$c1 = 30.6848629$
$c2 = 0.071243051$
$c3 = -0.00148796$

PUMP CURVE 2:

$P3 = P14 + c1 + c2*Q2 + c3*Q2^2$
$c1 = 30.6848629$
$c2 = 0.071243051$
$c3 = -0.00148796$

Status	Input	Name	Output	Unit
		v	0	
		a1	3.36899368	
		a2	-.059873165	
		T	50	DEG.F
		a3	.000515554	
		a4	-2.1467E-6	
		a5	3.430662E-9	
		rho	62.0031134	LB/CU.FT.
		b1	62.2614555	
		b2	.00122744	
		b3	-.000137041	
		b4	1.83106E-7	
		R1		
Guess	1	Q1		GPM
		D1	3	INCHES
		R2		
Guess	1	Q2		GPM
		D2	3	
		R3		
Guess	1	Q3		
		D3	3	
		R4		
Guess	1	Q4		
		D4	3	
		R5		
Guess	1	Q5		
		D5	3	
		R6		
Guess	1	Q6		
		D6	2	
		f1	108.396367	
		e	.00005	
		f2	-104.830604	
		f3	-39.8310993	
		f4	20.9499351	
		f5	-889.472793	
		f6	-55.5389961	
		V1	.124129018	FT./SEC.
		V2	.077961175	FT/SEC
		V3	.070004607	
		V4	.080690160	
		V5	.052873712	
		V6	.213247484	
		K1	.207960439	
		L1	15	FEET
		K2	-.201119456	
		L2	15	
		K3	-.101888935	
		L3	20	
		K4	.09378329	

Comment
WATER VISCOSITY AT TEMP T

WATER TEMPERATURE

WATER DENSITY AT TEMP. T

PIPE 1 REYNOLD'S NUMBER, TYPICAL
PUMP 1 DISCHARGE FLOW
PIPE 1 I.D., TYPICAL
PIPE 1 REYNOLDS NUMBER, TYPICAL
PUMP 2 DISCHARGE FLOW

PIPE 1 FRICTION FACTOR, TYPICAL
PIPE ROUGHNESS FACTOR

PIPE 1 FLOW VELOCITY, TYPICAL
PIPE 2 FLOW VELOCITY

PIPE 1 LENGTH, TYPICAL

Status	Input	Name	Output	Unit
		L4	35	
		K5	-4.3230574	
		L5	38	
		K6	-2.4813435	
		L6	46	
Guess	1	P1		FT.W.G.
		C1	30.6848629	
		C2	.071243051	
		C3	-.00148796	
Guess	1	P2		FT.W.G.
		K	2	
		KC	2	
		KG	.15	
		G	32.17	FT./SEC^2
Guess	1	P3		FT. W.G.
Guess	1	P5		
		KC1	16	
Guess	1	P4		
		KC2	16	
Guess	1	P6		FT. W.G.
Guess	1	P8		FT. W.G.
Guess	1	P7		
		R7		
Guess	1	Q7		
		D7	2	
		R8		
Guess	1	Q8		
		D8	2	
		R9		
Guess	1	Q9		
		D9	2	
		R10		
Guess	1	Q10		
		D10	2	
		R11		
Guess	1	Q11		
		D11	3	
		R12		
Guess	1	Q12		
		D12	3	
		R13		
Guess	1	Q13		
		D13	3	
		R14		
Guess	1	Q14		
		D14	3	
		R15		
Guess	1	Q15		
		D15	3	
		f7	-3.14650901	

Comment

PUMP 1 DISCHARGE PRESSURE
FACTOR FOR PUMP CURVE FIT

EXPONENT IN DARCY-WEISBACH DP EQUATION
K FACTOR FOR CHECK VALVE
K FACTOR FOR GATE VALVE
ACCELLERATION OF GRAVITY
PUMP 2 DISCHARGE PRESSURE

K FACTOR FOR COIL 1

K FACTOR FOR COIL 2

Status	Input	Name	Output	Unit
		f8	13.49445337	
		f9	29.04855276	
		f10	55.39507024	
		f11	-76.6372469	
		f12	268.294148	
		f13	-346.395343	
		f14		
		f15		
		V7	.287837225	
		V8	.289978587	
		V9	.235037538	
		V10	.204095785	
		V11	.084178873	
		V12	.050122786	
		V13	.050429968	
		V14	.110006864	
		V15	.080599651	
		K7	-.168082578	
		K8	.720856830	
		L8	55	
		L7	55	
		K9	1.26960330	
		L9	45	
		K10	2.47491325	
		L10	46	
		K11	-.372475943	
		L11	38	
		K12	1.20103034	
		L12	35	
		K13	-.88608784	
		L13	20	
		K14	-.459130547	
		L14	10	
		K15	9.81162958	
		L15	10	
		P13	5	FT. W.G.
Guess	1	P14		FT. W.G.
Guess	1	P10		
Guess	1	P9		
		KC3	10	
Guess	1	P11		
Guess	1	P12		
		PD1		PSI
		PD2		PSI

Comment

PUMP 1 SUCTION PRESSURE
PUMP 2 SUCTION PRESSURE

K FACTOR FOR COIL 3

PRESSURE RISE FOR PUMP 1
PRESSURE RISE FOR PUMP 2

Status	Input	Name	Output	Unit	Comment
		v	1.30736859:		WATER VISCOSITY AT TEMP T
		a1	.000557280:		
		a2	.22430157		
		T	50	DEG.F	WATER TEMPERATURE
		a3	-.0091107		
		a4	.000155423:		
		a5	-1.23904E-6		
		a6	3.790416E-9		
		rho	62.0031134	LB/CU.FT.	WATER DENSITY AT TEMP. T
		b1	62.2614555		
		b2	.001227444:		
		b3	-.000137041		
		b4	1.83106E-7		
		R1	167627.202		PIPE 1 REYNOLD'S NUMBER, TYPICAL
		Q1	139.703898:	GPM	PUMP 1 DISCHARGE FLOW
		D1	2	INCHES	PIPE 1 I.D., TYPICAL
		R2	47076.7759:		PIPE 1 REYNOLDS NUMBER, TYPICAL
		Q2	39.2347365(	GPM	PUMP 2 DISCHARGE FLOW
		D2	2		
		R3	111751.468		
		Q3	139.703898:		
		D3	3		
		R4	143135.985·		
		Q4	178.938634(		
		D4	3		
		R5	86876.0846(		
		Q5	108.606427:		
		D5	3		
		R6	49925.9257·		
		Q6	41.6092755(		
		D6	2		
		f1	.018145092		PIPE 1 FRICTION FACTOR, TYPICAL
		e	.00005		PIPE ROUGHNESS FACTOR
		f2	.022250978(		
		f3	.018725057(		
		f4	.018000795·		
		f5	.019546911:		

Variables

9.05 2 parallel pumps in network shacham 3 in SOL1.tkw

Status	Name	Output	Unit	Comment
	f6	.022004896		
	V1	14.2672606	FT./SEC.	PIPE 1 FLOW VELOCITY, TYPICAL
	V2	4.00684747		
	V3	6.34100471		
	V4	8.12182581		
	V5	4.92952506		
	V6	4.24934726		
	K1	.000264351		
	L1	15	FEET	PIPE 1 LENGTH, TYPICAL
	K2	.000324168		
	L2	15		
	K3	4.789916E-5		
	L3	20		
	K4	8.058134E-5		
	L4	35		
	K5	9.500282E-5		
	L5	38		
	K6	.000983123		
	L6	46		
	P1	50.3324556	FT.W.G.	PUMP 1 DISCHARGE PRESSURE
	C1	30.6848629		FACTOR FOR PUMP CURVE FIT
	C2	.071243051		
	C3	-.00148796		
	P2	38.3710281	FT.W.G.	
	K	2		EXPONENT IN DARCY-WEISBACH DP EQUATION
	KC	2		K FACTOR FOR CHECK VALVE
	KG	.15		K FACTOR FOR GATE VALVE
	G	32.17	FT./SEC^2	ACCELLERATION OF GRAVITY
	P3	38.4716578	FT. W.G.	PUMP 2 DISCHARGE PRESSURE
	P5	34.8560436		
	KC1	16		K FACTOR FOR COIL 1
	P4	37.4361713		
	KC2	16		K FACTOR FOR COIL 2
	P6	33.7354552	FT. W.G.	
	P8	13.0549815	FT. W.G.	
	P7	32.0333462		PUMP 1 SUCTION PRESSURE, GIVEN
	R7	84389.8511		

Status	Input	Name	Output	Unit	Comment
		Q7	70.3322075		
		D7	2		
		R8	80388.2012		
		Q8	66.9971517		
		D8	2		
		R9	49925.9257		
		Q9	41.6092755		
		D9	2		
		R10	49925.9257		
		Q10	41.6092755		
		D10	2		
		R11	86876.0846		
		Q11	108.606427		
		D11	3		
		R12	143135.985		
		Q12	178.938634		
		D12	3		
		R13	111751.468		
		Q13	139.703898		
		D13	3		
		R14	111751.468		
		Q14	139.703898		
		D14	3		
		R15	47076.7759		
		Q15	39.2347365		
		D15	2		
		f7	.020061642		
		f8	.020223079		
		f9	.022004896		
		f10	.022004896		
		f11	.019546911		
		f12	.018000795		
		f13	.018725057		
		f14	.018725057		
		f15	.022250978		
		V7	7.18267669		
		V8	6.84208411		

Variables

9.05 2 parallel pumps in network shacham 3 in SOL1.tkw

Status	Input	Name	Output	Unit	Comment
		V9	4.24934726		
		V10	4.24934726		
		V11	4.92952506		
		V12	8.12182581		
		V13	6.34100471		
		V14	6.34100471		
		V15	4.00684747		
		K7	.0010716667		
		K8	.0010802910		
		L8	55		
		L7	55		
		K9	.000961751		
		L9	45		
		K10	.000983123		
		L10	46		
		K11	9.500282E-5		
		L11	38		
		K12	8.058134E-5		
		L12	35		
		K13	4.789916E-5		
		L13	20		
		K14	2.394958E-5		
		L14	10		
		K15	.000216112		
		L15	10		
		P13	5	FT. W.G.	PUMP 1 SUCTION PRESSURE
		P14	7.28210552	FT. W.G.	PUMP 2 SUCTION PRESSURE
		P10	10.2322841		
		P9	11.3528725		
		KC3	10		K FACTOR FOR COIL 3
		P11	7.65215645		
		P12	6.71729965		

Variables   9.05 2 parallel pumps in network shacham 3 in SOL1.tkw

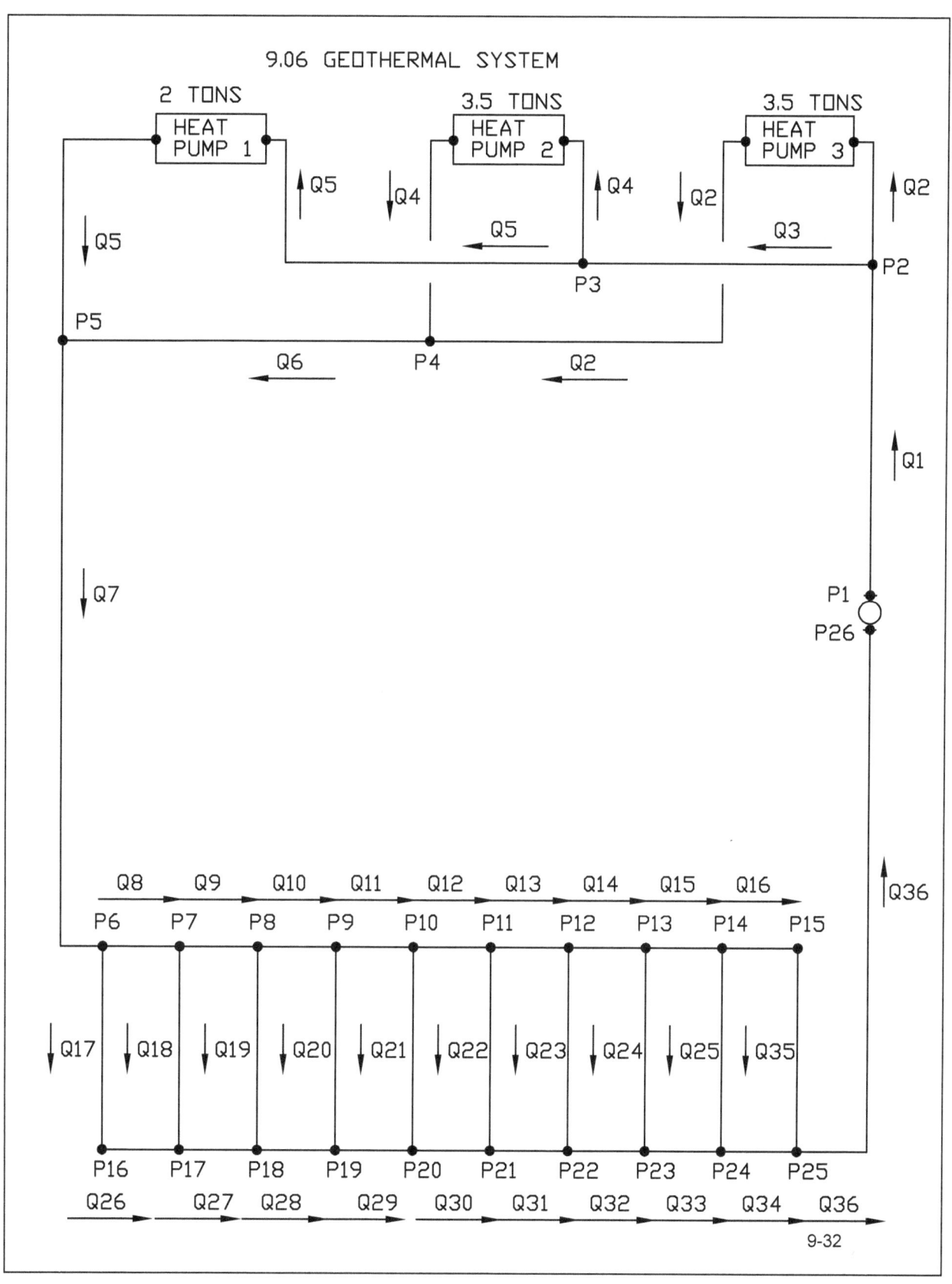

Rule

$v = a1 + a2*T + a3*T^2 + a4*T^3 + a5*T^4 + a6*T^5$

$rho = b1 + b2*T + b3*T^2 + b4*T^3$

R1=50.6*Q1*rho/(D1*v)
R2=50.6*Q3*rho/(D2*v)
R3=50.6*Q3*rho/(D3*v)
R4=50.6*Q4*rho/(D4*v)
R5=50.6*Q5*rho/(D5*v)
R6=50.6*Q6*rho/(D6*v)
R7=50.6*Q7*rho/(D7*v)
R8=50.6*Q8*rho/(D8*v)
R9=50.6*Q9*rho/(D9*v)
R10=50.6*Q10*rho/(D10*v)
R11=50.6*Q11*rho/(D11*v)
R12=50.6*Q12*rho/(D12*v)
R13=50.6*Q13*rho/(D13*v)
R14=50.6*Q14*rho/(D14*v)
R15=50.6*Q15*rho/(D15*v)
R16=50.6*Q16*rho/(D16*v)
R17=50.6*Q17*rho/(D17*v)
R18=50.6*Q18*rho/(D18*v)
R19=50.6*Q19*rho/(D19*v)
R20=50.6*Q20*rho/(D20*v)
R21=50.6*Q21*rho/(D21*v)
R22=50.6*Q22*rho/(D22*v)
R23=50.6*Q23*rho/(D23*v)
R24=50.6*Q24*rho/(D24*v)
R25=50.6*Q25*rho/(D25*v)
R26=50.6*Q26*rho/(D26*v)
R27=50.6*Q27*rho/(D27*v)
R28=50.6*Q28*rho/(D28*v)
R29=50.6*Q29*rho/(D29*v)
R30=50.6*Q30*rho/(D30*v)
R31=50.6*Q31*rho/(D31*v)
R32=50.6*Q32*rho/(D32*v)
R33=50.6*Q33*rho/(D33*v)
R34=50.6*Q34*rho/(D34*v)
R35=50.6*Q35*rho/(D35*v)
R36=50.6*Q36*rho/(D36*v)

f1=(-2*log((e*12/D1)/3.7-(5.02/R1)*log((e*12/D1)/3.7+14.5/R1)))^-2
f2=(-2*log((e*12/D2)/3.7-(5.02/R2)*log((e*12/D2)/3.7+14.5/R2)))^-2
f3=(-2*log((e*12/D3)/3.7-(5.02/R3)*log((e*12/D3)/3.7+14.5/R3)))^-2
f4=(-2*log((e*12/D4)/3.7-(5.02/R4)*log((e*12/D4)/3.7+14.5/R4)))^-2
f5=(-2*log((e*12/D5)/3.7-(5.02/R5)*log((e*12/D5)/3.7+14.5/R5)))^-2
f6=(-2*log((e*12/D6)/3.7-(5.02/R6)*log((e*12/D6)/3.7+14.5/R6)))^-2
f7=(-2*log((e*12/D7)/3.7-(5.02/R7)*log((e/D7)/3.7+14.5/R7)))^-2
f8=(-2*log((e*12/D8)/3.7-(5.02/R8)*log((e*12/D8)/3.7+14.5/R8)))^-2
f9=(-2*log((e*12/D9)/3.7-(5.02/R9)*log((e*12/D9)/3.7+14.5/R9)))^-2

Rule
f10=(-2*log((e*12/D10)/3.7-(5.02/R10)*log((e*12/D10)/3.7+14.5/R10)))^-2
f11=(-2*log((e*12/D11)/3.7-(5.02/R11)*log((e*12/D11)/3.7+14.5/R11)))^-2
f12=(-2*log((e*12/D12)/3.7-(5.02/R12)*log((e*12/D12)/3.7+14.5/R12)))^-2
f13=(-2*log((e*12/D13)/3.7-(5.02/R13)*log((e*12/D13)/3.7+14.5/R13)))^-2
f14=(-2*log((e*12/D14)/3.7-(5.02/R14)*log((e*12/D14)/3.7+14.5/R14)))^-2
f15=(-2*log((e*12/D15)/3.7-(5.02/R15)*log((e*12/D15)/3.7+14.5/R15)))^-2
f16=(-2*log((e*12/D16)/3.7-(5.02/R16)*log((e*12/D16)/3.7+14.5/R16)))^-2
f17=(-2*log((e*12/D17)/3.7-(5.02/R17)*log((e*12/D17)/3.7+14.5/R17)))^-2
f18=(-2*log((e*12/D18)/3.7-(5.02/R18)*log((e*12/D18)/3.7+14.5/R18)))^-2
f19=(-2*log((e*12/D19)/3.7-(5.02/R19)*log((e*12/D19)/3.7+14.5/R19)))^-2
f20=(-2*log((e*12/D20)/3.7-(5.02/R20)*log((e*12/D20)/3.7+14.5/R20)))^-2
f21=(-2*log((e*12/D21)/3.7-(5.02/R21)*log((e*12/D21)/3.7+14.5/R21)))^-2
f22=(-2*log((e*12/D22)/3.7-(5.02/R22)*log((e*12/D22)/3.7+14.5/R22)))^-2
f23=(-2*log((e*12/D23)/3.7-(5.02/R23)*log((e*12/D23)/3.7+14.5/R23)))^-2
f24=(-2*log((e*12/D24)/3.7-(5.02/R24)*log((e*12/D24)/3.7+14.5/R24)))^-2
f25=(-2*log((e*12/D25)/3.7-(5.02/R25)*log((e*12/D25)/3.7+14.5/R25)))^-2
f26=(-2*log((e*12/D26)/3.7-(5.02/R26)*log((e*12/D26)/3.7+14.5/R26)))^-2
f27=(-2*log((e*12/D27)/3.7-(5.02/R27)*log((e*12/D27)/3.7+14.5/R27)))^-2
f28=(-2*log((e*12/D28)/3.7-(5.02/R28)*log((e*12/D28)/3.7+14.5/R28)))^-2
f29=(-2*log((e*12/D29)/3.7-(5.02/R29)*log((e*12/D29)/3.7+14.5/R29)))^-2
f30=(-2*log((e*12/D30)/3.7-(5.02/R30)*log((e*12/D30)/3.7+14.5/R30)))^-2
f31=(-2*log((e*12/D31)/3.7-(5.02/R31)*log((e*12/D31)/3.7+14.5/R31)))^-2
f32=(-2*log((e*12/D32)/3.7-(5.02/R32)*log((e*12/D32)/3.7+14.5/R32)))^-2
f33=(-2*log((e*12/D33)/3.7-(5.02/R33)*log((e*12/D33)/3.7+14.5/R33)))^-2
f34=(-2*log((e*12/D34)/3.7-(5.02/R34)*log((e*12/D34)/3.7+14.5/R34)))^-2
f35=(-2*log((e*12/D35)/3.7-(5.02/R35)*log((e*12/D35)/3.7+14.5/R35)))^-2
f36=(-2*log((e*12/D36)/3.7-(5.02/R36)*log((e*12/D36)/3.7+14.5/R36)))^-2

VEL1=.4085*Q1/(D1^2)
VEL2=.4085*Q2/(D2^2)
VEL3=.4085*Q3/(D3^2)
VEL4=.4085*Q4/(D4^2)
VEL5=.4085*Q5/(D5^2)
VEL6=.4085*Q6/(D6^2)
VEL7=.4085*Q7/(D7^2)
VEL8=.4085*Q8/(D8^2)
VEL9=.4085*Q9/(D9^2)
VEL10=.4085*Q10/(D10^2)

C1=0.03108*f1*L1/(D1^5)
C2=0.03108*f2*L2/(D2^5)
C3=0.03108*f3*L3/(D3^5)
C4=0.03108*f4*L4/(D4^5)
C5=0.03108*f5*L5/(D5^5)
C6=0.03108*f6*L6/(D6^5)
C7=0.03108*f7*L7/(D7^5)
C8=0.03108*f8*L8/(D8^5)
C9=0.03108*f9*L9/(D9^5)
C10=0.03108*f10*L10/(D10^5)
C11=0.03108*f11*L11/(D11^5)

Rule
C12=0.03108*f12*L12/(D12^5)
C13=0.03108*f13*L13/(D13^5)
C14=0.03108*f14*L14/(D14^5)
C15=0.03108*f15*L15/(D15^5)
C16=0.03108*f16*L16/(D16^5)
C17 =0.03108*f17*L17/(D17^5)
C18=0.03108*f18*L18/(D18^5)
C19=0.03108*f19*L19/(D19^5)
C20=0.03108*f20*L20/(D20^5)
C21=0.03108*f21*L21/(D21^5)
C22=0.03108*f22*L22/(D22^5)
C23=0.03108*f23*L23/(D23^5)
C24=0.03108*f24*L24/(D24^5)
C25=0.03108*f25*L25/(D25^5)
C26=0.03108*f26*L26/(D26^5)
C27=0.03108*f27*L27/(D27^5)
C28=0.03108*f28*L28/(D28^5)
C29=0.03108*f29*L29/(D29^5)
C30=0.03108*f30*L30/(D30^5)
C31=0.03108*f31*L31/(D31^5)
C32=0.03108*f32*L32/(D32^5)
C33=0.03108*f33*L33/(D33^5)
C34=0.03108*f34*L34/(D34^5)
C35=0.03108*f35*L35/(D35^5)
C36=0.03108*f36*L36/(D36^5)

P1-P2=C1*Q1^2
P2-P4=C2*Q2^2+2.2-.566*Q2+.111*Q2^2
P2-P3=C3*Q3^2
P3-P5=C5*Q5^2+2.5-1.166*Q5+.422*Q5^2
P3-P4=C4*Q4^2+2.2-.566*Q4+.111*Q4^2
P4-P5=C6*Q6^2
P5-P6=C7*Q7^2
P6-P7=C8*Q8^2
P7-P8=C9*Q9^2
P8-P9=C10*Q10^2
P9-P10=C11*Q11^2
P10-P11=C12*Q12^2
P11-P12=C13*Q13^2
P12-P13=C14*Q14^2
P13-P14=C15*Q15^2
P14-P15=C16*Q16^2
P15-P25=C35*Q35^2
P6-P16=C17*Q17^2
P16-P17=C26*Q26^2
P7-P17=C18*Q18^2
P17-P18=C27*Q27^2
P8-P18=C19*Q19^2
P18-P19=C28*Q28^2
P9-P19=C20*Q20^2

Rule

$P_{19}-P_{20}=C_{29}*Q_{29}^2$
$P_{10}-P_{20}=C_{21}*Q_{21}^2$
$P_{20}-P_{21}=C_{30}*Q_{30}^2$
$P_{11}-P_{21}=C_{22}*Q_{22}^2$
$P_{21}-P_{22}=C_{31}*Q_{31}^2$
$P_{12}-P_{22}=C_{23}*Q_{23}^2$
$P_{22}-P_{23}=C_{32}*Q_{32}^2$
$P_{13}-P_{23}=C_{24}*Q_{24}^2$
$P_{23}-P_{24}=C_{33}*Q_{33}^2$
$P_{14}-P_{24}=C_{25}*Q_{25}^2$
$P_{24}-P_{25}=C_{34}*Q_{34}^2$
$P_{25}-P_{26}=C_{36}*Q_{36}^2$

$Q_1=Q_2+Q_3$
$Q_3=Q_4+Q_5$
$Q_2+Q_4=Q_6$
$Q_5+Q_6=Q_7$
$Q_1=Q_7$
$Q_7=Q_8+Q_{17}$
$Q_8=Q_9+Q_{18}$
$Q_9=Q_{10}+Q_{19}$
$Q_{10}=Q_{11}+Q_{20}$
$Q_{11}=Q_{12}+Q_{21}$
$Q_{12}=Q_{13}+Q_{22}$
$Q_{13}=Q_{14}+Q_{23}$
$Q_{14}=Q_{15}+Q_{24}$
$Q_{15}=Q_{16}+Q_{25}$
$Q_{16}=Q_{35}$
$Q_{17}=Q_{26}$
$Q_{26}+Q_{18}=Q_{27}$
$Q_{27}+Q_{19}=Q_{28}$
$Q_{28}+Q_{20}=Q_{29}$
$Q_{29}+Q_{21}=Q_{30}$
$Q_{30}+Q_{22}=Q_{31}$
$Q_{31}+Q_{23}=Q_{32}$
$Q_{32}+Q_{24}=Q_{33}$
$Q_{33}+Q_{25}=Q_{34}$
$Q_{34}+Q_{35}=Q_{36}$

$P_1-P_{26}=PHEAD$

Status	Input	Name	Output	Unit	Comment
		v	.7573792930		WATER VISCOSITY AT TEMP T
	.0005572800	a1			
	.22430157	a2			
	90	T		DEG. F	AVERAGE WATER TEMP.
	-.0091107	a3			
	.000155423	a4			
	-1.23944E-6	a5			
	3.790416E-9	a6			
		rho	62.1254802		
	62.2614555	b1			
	.009339695	b2			
	-.000137041	b3			
	1.83106E-7	b4			
		R1	60623.7457		
	28	Q1		GPM	TOTAL FLOW, GIVEN
	1.917	D1			
		R2	45571.2292		
		Q2	16.8426046		
	1.534	D2			
		R3	45571.2292		
	1.534	D3			
		R4	29888.1760		
		Q4	11.0463277	GPM	FLOW THROUGH HEAT PUMP 2
	1.534	D4			
		R5	15683.0531		
		Q5	5.79627691	GPM	FLOW THROUGH HEAT PUMP 1
	1.534	D5			
		R6	60076.8690		
		Q6	22.2037230		
	1.534	D6			
		R7	60623.7457		
		Q7	28		
	1.917	D7			
		R8	55104.6707		
		Q8	25.4509311		
	1.917	D8			
		R9	49717.1368		

Status	Input	Name	Output	Unit	Comment
		Q9	22.9626166		
	1.917	D9			
		R10	44242.8693		
		Q10	20.4342428		
	1.917	D10			
	.00015	e			
		f1	.0232706956		FRICTION FACTOR, PIPE 1, TYPICAL
		f2	.024809846		
		f3	.024809846		
		f4	.026400963		
		f5	.029644188		
		f6	.023963862		
		f7	.023535621		
		f8	.023565085		
		f9	.023901497		
		f10	.024307749		
		VEL1	3.11247496		
		VEL2	1.93688646		
		Q2	11.1573953	GPM	FLOW THROUGH HEAT PUMP 3
		VEL3	2.92381975		
		VEL4	1.91760549		
		VEL5	1.00621425		
		VEL6	3.85449195		
		VEL7	3.11247496		
		VEL8	2.82912093		
		VEL9	2.55252033		
		VEL10	2.27146675		
		C1	.00027937		
	10	L1			
		C2	.000907775		
	10	L2			
		C3	.000726220		
	8	L3			
		C4	.000965993		
	10	L4			
		C5	.001084660		
	10	L5			

Variables

dw shacham 35 pipes geothermal.tkw

Variables

Status	Input	Name	Output	Unit	Comment
		C6	.000701456		
	8	L6			
		C7	.005651010		
	200	L7			
		C8	.000424356		
	15	L8			
		C9	.000430414		
	15	L9			
		C10	.000437730		
	15	L10			
		R11	48536.5329		
		f11	.024603936		
	1.534	D11			
		R12	42057.8499		
		f12	.025083436		
	1.534	D12			
		R13	35689.2606		
		f13	.025685056		
	1.534	D13			
		R14	29262.7457		
		f14	.026491062		
	1.534	D14			
		R15	34618.3465		
		f15	.027176985		
	1.002	D15			
		R16	20012.2424		
		f16	.029429267		
	1.002	D16			
		R17	10558.9488		
		f17	.033090228		
	1.002	D17			
		R18	10307.2879		
		f18	.033253213		
	1.002	D18			
		R19	10473.2244		
		f19	.033145089		
	1.002	D19			

dw shacham 35 pipes geothermal.tkw

Status	Input	Name	Output	Unit	Comment
		R20	10337.8631		
	1.002	f20	.033233095		
		D20			
		R25	14606.1041		
	1.002	f25	.031079649		
		D25			
		R21	9918.46277		
	1.002	f21	.033517055		
		D21			
		R22	9749.91622		
	1.002	f22	.033636265		
		D22			
		R23	9838.59659		
	1.002	f23	.033573164		
		D23			
		R24	10181.1064		
	1.002	f24	.033337186		
		D24			
		R26	10558.9488		
	1.002	f26	.033090228		
		D26			
		R27	20866.2368		
	1.002	f27	.029230927		
		D27			
		R28	20470.7563		
	1.534	f28	.028175476		
		D28			
		R29	27223.3892		
	1.534	f29	.026806681		
		D29			
		R30	33702.0722		
	1.534	f30	.025908603		
		D30			
		C11	.001350361		
	15	L11			
		C12	.001376678		
	15	L12			

Variables

Status	Input	Name	Output	Unit	Comment
		C13	.001409697		
	15	L13			
		C14	.0014539345		
	15	L14			
		C15	.012543968		
	15	L15			
		C16	.013583544		
	15	L16			
		C17	.637406527		
	626	L17			
		C18	.640546038		
	626	L18			
		C19	.638463293		
	626	L19			
		C20	.640158526		
	626	L20			
		C21	.645628349		
	626	L21			
		C22	.647924652		
	626	L22			
		C23	.646709156		
	626	L23			
		C24	.642163591		
	626	L24			
		C25	.598677376		
	626	L25			
		C26	.015273319		
	15	L26			
		C27	.013491997		
	15	L27			
		C28	.0015463815		
	15	L28			
		C29	.001471257		
	15	L29			
		C30	.0014219665		
	15	L30			
		C31	.001386087		

dw shacham 35 pipes geothermal.tkw

Status	Input	Name	Output	Unit	Comment
		f31	.025254873		
	15	L31			
	1.534	D31			
		C32	.001358009		
		f32	.024743288		
	15	L32			
	1.534	D32			
		C33	.011508558		
		f33	.024933730		
	15	L33			
	1.002	D33			
		C34	.011373025		
		f34	.024640092		
	15	L34			
	1.002	D34			
		C35	.566886600		
		f35	.029429267		
	626	L35			
	1.002	D35			
		R31	40070.6615		
		R32	46497.1764		
		R33	81365.4065		
		R34	95971.5106		
		R35	20012.2424		
		Q11	17.9385469		
		Q12	15.5440999		
		Q13	13.1903421		
		Q14	10.8151758		
		Q15	8.35732314		
		Q16	4.83121793		
		Q17	2.54906882		
		Q18	2.48831456		
		Q19	2.52837381		
		Q20	2.49569582		
		Q21	2.39444706		
		Q22	2.35375772		
		Q23	2.37516632		

Variables

dw shacham 35 pipes geothermal.tkw

Status	Input	Name	Output	Unit	Comment
		Q24	2.457852709		
		Q25	3.526105211		
		Q26	2.549068821		
		Q27	5.037383380		
		Q28	7.565757191		
		Q29	10.06145301		
		Q30	12.4559001		
		Q31	14.80965781		
		Q32	17.18482411		
		Q33	19.64267681		
		Q34	23.16878201		
		Q35	4.831217931		
		R36	60623.74571		
		Q36	28		
	1.917	D36			
		f36	.0232706956		
		C36	.0055874006		
	200	L36			
		P1	45.48358549		
		P2	45.26455931		
		P4	35.44852921		
		P3	45.0585501		
		P5	35.10270721		
		P6	30.67231491		
		P7	30.3974382		
		P8	30.17048851		
		P9	29.98771061		
		P10	29.55317581		
		P11	29.22054411		
		P12	28.97527761		
		P13	28.80521381		
		P14	27.92908221		
		P15	27.61203321		
		P25	14.38052201		
		P16	26.53060551		
		P17	26.43136321		
		P18	26.08900071		

Status	Input	Name	Output	Unit	Comment
		P19	26.0004847		
		P20	25.8515452		
		P21	25.6309278		
		P22	25.3269229		
		P23	24.9258779		
		P24	20.4854762		
	10	P26		FT. W.G.	PRESSURE TO PUMP, GIVEN
		PHEAD	35.4835854	FT. W.G.	HEAD REQUIRED FOR PUMP

Variables

dw shacham 35 pipes geothermal.tkw

Submittal Data Information 121-138 Series Circulators

FIGURE 9.15 TACO 133 PUMP CURVE

Use this curve

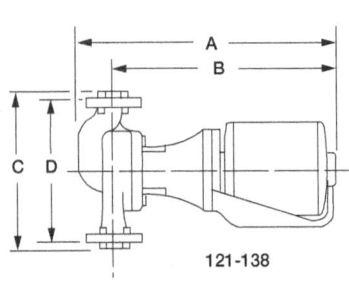

121-138

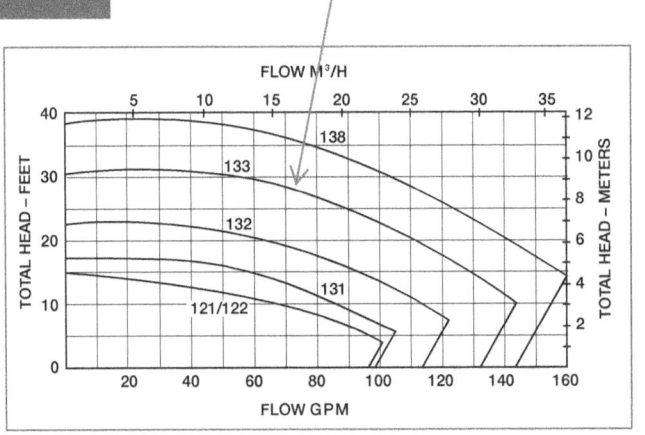

Specifications (121-138 Series)

MODEL NUMBER[1]	FLANGE SIZE	MOTOR[2] 60C/AC 1 PH	RPM	DIMENSIONS								SHIPPING WEIGHT	
				A		B		C		D			
				IN.	MM	IN.	MM	IN.	MM	IN.	MM	LBS.	KG
121	2½"	¼ HP, 115 V	1725	18⅛	460.4	15⅞	403.2	14¼	362.0	11⅛	282.6	72	32.7
122	3"	¼ HP, 115 V	1725	18⅛	460.4	15⅞	403.2	13⅝	346.1	11⅛	282.6	72	32.7
131	3"	⅓ HP, 115 V	1725	19¼	489.0	15¾	400.1	16	406.4	13⅝	346.1	95	43.1
132	3"	½ HP, 115/230 V or 230/460/60/3 or 200/60/3	1725	21½	546.1	18	457.2	16	406.4	13⅝	346.1	108	49.0
133	3"	¾ HP, 115/230 V or 230/460/60/3 or 200/60/3	1725	22⅛	562.0	18⅝	473.1	16	406.4	13⅝	346.1	113	51.3
138	3"	1 HP, 115/230 V or 230/460/60/3 or 200/60/3	1725	22⅝	574.0	19⅛	485.8	16	406.4	13⅝	346.1	118	53.5

(1) When specifying all bronze construction, add letter "B" after model number (i.e. 132B).
When specifying bronze fitted construction, add letter "C" after model number (i.e. 132C).
(2) Motors are available with other electrical characteristics – consult your Taco representative.

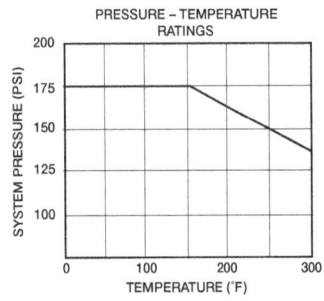

FIGURE 9.1

Taco Inc., 1160 Cranston Street, Cranston, RI 02920 / (401) 942-8000 / Fax (401) 942-2360
Taco (Canada) Ltd., 8450 Lawson Road, Unit #3, Milton, Ontario L9T 0J8 / (905) 564-9422 / Fax (905) 564-9436
www.taco-hvac.com

FIGURE 9.16 TACO 133 CURVE FIT

Polynomial Regression Worksheet $(y = b_1 + b_2 \cdot x + b_3 \cdot x^2 + ...)$

	X	Y	residuals	SUMMARY STATS	
1	0	30.9	+2.151E-1	order	2
2	10	31.2	-4.849E-2	N	15
3	20	31.4	-1.145E-1	Syx	.1623296266
4	30	31.3	-1.830E-1	adj R2	.9994027952
5	40	31	-1.538E-1	p	6.943195E-8
6	50	30.5	-2.710E-2		
7	60	29.8	+1.972E-1	b1	30.6848529
8	70	28.5	+1.191E-1	b2	.071243051
9	80	27	+1.387E-1	b3	-.001487961
10	90	25	-4.425E-2	b4	0
11	100	23	+7.045E-2	b5	0
12	110	20.5	-1.727E-2	b6	0
13	120	17.5	-3.074E-1	b7	0
14	130	14.95	+1.501E-1	b8	0
15	140	11.5	+5.147E-3	b9	0

INT TABLE: works   taco 133.tkw

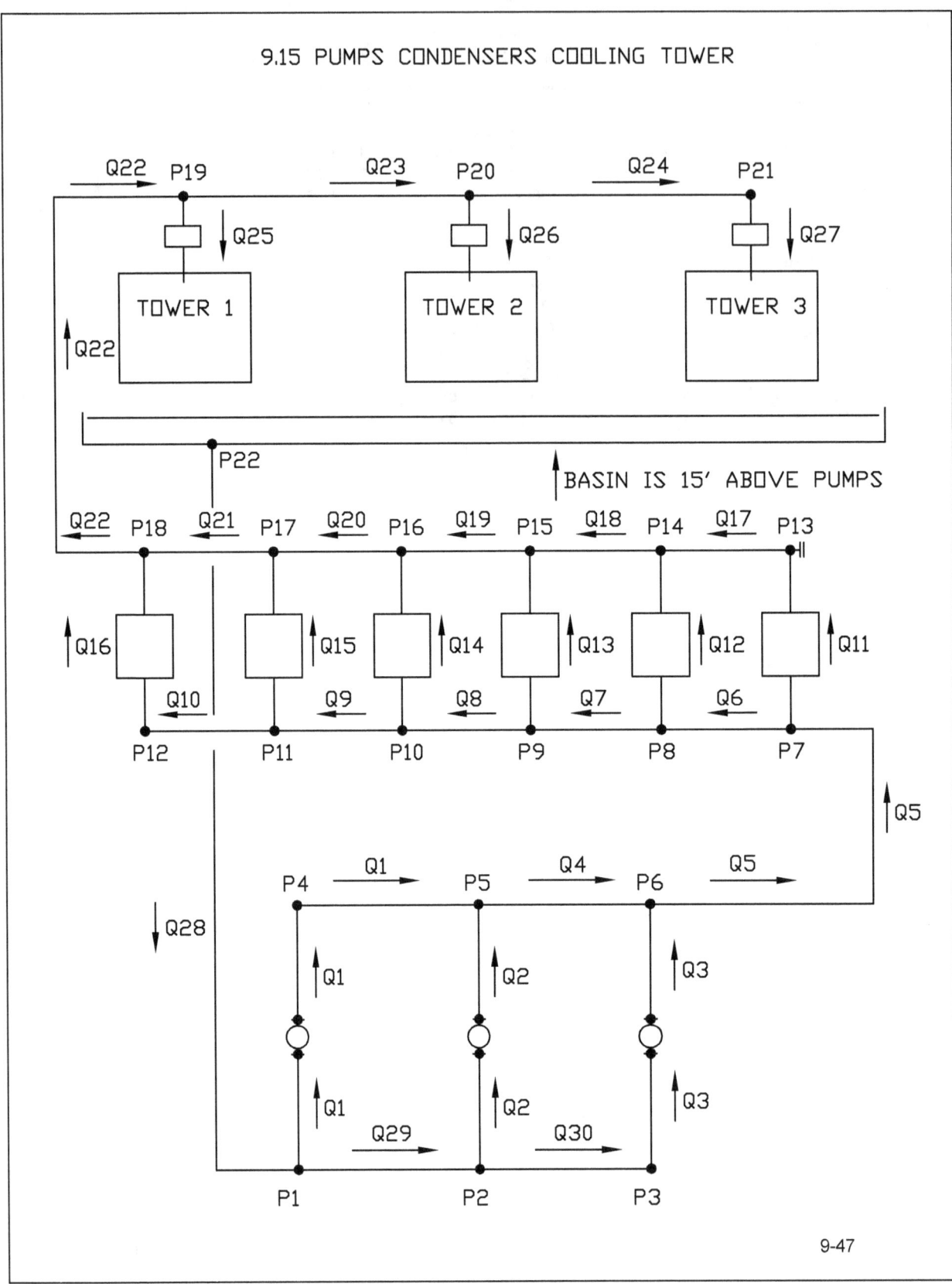

Rule

;>>>/* TK SOLVER */
;>>>/* HAZEN WILLIAMS, FT.W.G. */
;>>>/* PCCT WITH PUMPS */
;>>>/* THIS GOES FROM SUCTION OF 1ST PUMP TO THE LAST COOLING TOWER */

K1=0.002083*L1*(100/C)^1.85/(D1^4.8655)
K4=0.002083*L4*(100/C)^1.85/(D4^4.8655)
K5=0.002083*L5*(100/C)^1.85/(D5^4.8655)
K6=0.002083*L6*(100/C)^1.85/(D6^4.8655)
K7=0.002083*L7*(100/C)^1.85/(D7^4.8655)
K8=0.002083*L8*(100/C)^1.85/(D8^4.8655)
K9=0.002083*L9*(100/C)^1.85/(D9^4.8655)
K10=0.002083*L10*(100/C)^1.85/(D10^4.8655)
K11=0.002083*L11*(100/C)^1.85/(D11^4.8655)
K12=0.002083*L12*(100/C)^1.85/(D12^4.8655)
K13=0.002083*L13*(100/C)^1.85/(D13^4.8655)
K14=0.002083*L14*(100/C)^1.85/(D14^4.8655)
K15=0.002083*L15*(100/C)^1.85/(D15^4.8655)
K16=0.002083*L16*(100/C)^1.85/(D16^4.8655)
K17=0.002083*L17*(100/C)^1.85/(D17^4.8655)
K18=0.002083*L18*(100/C)^1.85/(D18^4.8655)
K19=0.002083*L19*(100/C)^1.85/(D19^4.8655)
K20=0.002083*L20*(100/C)^1.85/(D20^4.8655)
K21=0.002083*L21*(100/C)^1.85/(D21^4.8655)
K22=0.002083*L22*(100/C)^1.85/(D22^4.8655)
K23=0.002083*L23*(100/C)^1.85/(D23^4.8655)
K24=0.002083*L24*(100/C)^1.85/(D24^4.8655)
K28=0.002083*L28*(100/C)^1.85/(D28^4.8655)
K29=0.002083*L29*(100/C)^1.85/(D29^4.8655)
K30=0.002083*L30*(100/C)^1.85/(D30^4.8655)

V1=0.4085*Q1/(D1^2)
V4=0.4085*Q4/(D4^2)
V5=0.4085*Q5/(D5^2)
V6=0.4085*Q6/(D6^2)
V7=0.4085*Q7/(D7^2)
V8=0.4085*Q8/(D8^2)
V9=0.4085*Q9/(D9^2)
V10=0.4085*Q10/(D10^2)
V11=0.4085*Q11/(D11^2)
V12=0.4085*Q12/(D12^2)
V13=0.4085*Q13/(D13^2)
V14=0.4085*Q14/(D14^2)
V15=0.4085*Q15/(D15^2)
V16=0.4085*Q16/(D16^2)
V17=0.4085*Q17/(D1^2)
V18=0.4085*Q18/(D18^2)
V19=0.4085*Q19/(D19^2)
V20=0.4085*Q20/(D20^2)

Rule

$V21 = 0.4085 \cdot Q21/(D21^2)$
$V22 = 0.4085 \cdot Q22/(D22^2)$
$V23 = 0.4085 \cdot Q23/(D23^2)$
$V24 = 0.4085 \cdot Q24/(D24^2)$
$V29 = 0.4085 \cdot Q29/(D29^2)$
$V30 = 0.4085 \cdot Q30/(D30^2)$

;>>>/* LINK EQUATIONS */
P6-P7=K5*Q5^K
P7-P8=K6*Q6^K
P8-P9=K7*Q7^K
P9-P10=K8*Q8^K
P10-P11=K9*Q9^K
P11-P12=K10*Q10^K
P13-P14=K17*Q17^K
P14-P15=K18*Q18^K
P15-P16=K19*Q19^K
P16-P17=K20*Q20^K
P17-P18=K21*Q21^K
P7-P13=K11*Q11^K+(Q11/KCOND)^2
P8-P14=K12*Q12^K+(Q12/KCOND)^2
P9-P15=K13*Q13^K+(Q13/KCOND)^2
P10-P16=K14*Q14^K+(Q14/KCOND)^2
P11-P17=K15*Q15^K+(Q15/KCOND)^2
P12-P18=K16*Q16^K+(Q16/KCOND)^2
P18-P19=K22*Q22^K+15
P19-P20=K23*Q23^K
P20-P21=K24*Q24^K
P19=(Q25/KTWR)^2
P20=(Q26/KTWR)^2
P21=(Q27/KTWR)^2

P4=P1+69.619+.00300793*Q1-6.8095E-6*Q1^2+4.444E-10*Q1^3
P5=P2+69.619+.00300793*Q2-6.8095E-6*Q2^2+4.444E-10*Q2^3
P6=P3+69.619+.00300793*Q3-6.8095E-6*Q3^2+4.444E-10*Q3^3

P1-P2=K29*Q29^K
P2-P3=K30*Q30^K

P4-P5=K1*Q1^K
P5-P6=K4*Q4^K

;>>>/* NODAL EQUATIONS */
Q1+Q2=Q4
Q3+Q4=Q5

Q5=Q1+Q29
Q29=Q2+Q30
Q30=Q3

Rule
Q5=Q11+Q6
Q6=Q7+Q12
Q7=Q13+Q8
Q8=Q14+Q9
Q9=Q15+Q10
Q10=Q16
Q11=Q17
Q12+Q17=Q18
Q13+Q18=Q19
Q14+Q19=Q20
Q15+Q20=Q21
Q16+Q21=Q22
Q22=Q25+Q23
Q23=Q26+Q24
Q24=Q27

;>>>/* GIVEN */
K=1.85
C=130
KCOND=258.2
KTWR=632.45
P1=15

;>>>/* PIPE LENGTHS AND DIAMETERS */
L1=10
D1=8
L4=6
D4=12
L5=55
D5=12
L6=10
D6=12
L7=10
D7=10
L8=10
D8=10
L9=10
D9=8
L10=10
D10=6
L11=10
D11=6
L12=10
D12=6
L13=10
D13=6
L14=10
D14=6
L15=10
D15=6

Rule
L16=10
D16=6
L17=10
D17=6
L18=10
D18=8
L19=10
D19=10
L20=10
D20=10
L21=10
D21=12
L22=25
D22=12
L23=15
D23=10
L24=15
D24=8
L28=25
D28=12
L29=5
D29=12
L30=5
D30=8

C = 130.0
D1 = 8.0
D4 = 12.0
D5 = 12.0
D6 = 12.0
D7 = 10.0
D8 = 10.0
D9 = 8.0
D10 = 6.0
D11 = 6.0
D12 = 6.0
D13 = 6.0
D14 = 6.0
D15 = 6.0
D16 = 6.0
D17 = 6.0
D18 = 8.0
D19 = 10.0
D20 = 10.0
D21 = 12.0
D22 = 12.0
D23 = 10.0
D24 = 8.0
D28 = 12.0
D29 = 12.0

Rule
D30 = 8.0
K = 1.85
KCOND = 258.2
KTWR = 632.45
K1 = 5.174992640066056E-7
K4 = 4.318062916206087E-8
K5 = 3.958224339855579E-7
K6 = 7.196771527010144E-8
K7 = 1.7474069881108927E-7
K8 = 1.7474069881108927E-7
K9 = 5.174992640066056E-7
K10 = 2.0979694480700837E-6
K11 = 2.0979694480700837E-6
K12 = 2.0979694480700837E-6
K13 = 2.0979694480700837E-6
K14 = 2.0979694480700837E-6
K15 = 2.0979694480700837E-6
K16 = 2.0979694480700837E-6
K17 = 2.0979694480700837E-6
K18 = 5.174992640066056E-7
K19 = 1.7474069881108927E-7
K20 = 1.7474069881108927E-7
K21 = 7.196771527010144E-8
K22 = 1.799192881752536E-7
K23 = 2.621110482166339E-7
K24 = 7.762488960099083E-7
K28 = 1.799192881752536E-7
K29 = 3.598385763505072E-8
K30 = 2.587496320033028E-7
L1 = 10.0
L4 = 6.0
L5 = 55.0
L6 = 10.0
L7 = 10.0
L8 = 10.0
L9 = 10.0
L10 = 10.0
L11 = 10.0
L12 = 10.0
L13 = 10.0
L14 = 10.0
L15 = 10.0
L16 = 10.0
L17 = 10.0
L18 = 10.0
L19 = 10.0
L20 = 10.0
L21 = 10.0
L22 = 25.0
L23 = 15.0

Rule
L24 = 15.0
L28 = 25.0
L29 = 5.0
L30 = 5.0

FIG. 9.19 PCCT SOLUTION

Status	Input	Name	Output	Unit	Comment
		K1	5.174993E-		
		L1	10		
		C	130		
		D1	8		
		K4	4.318063E-8		
		L4	6		
		D4	12		
		K5	3.958224E-		
		L5	55		
		D5	12		
		K6	7.196772E-8		
		L6	10		
		D6	12		
		K7	1.747407E-		
		L7	10		
		D7	10		
		K8	1.747407E-		
		L8	10		
		D8	10		
		K9	5.174993E-		
		L9	10		
		D9	8		
		K10	2.097969E-6		
		L10	10		
		D10	6		
		K11	2.097969E-6		
		L11	10		
		D11	6		
		K12	2.097969E-6		
		L12	10		
		D12	6		
		K13	2.097969E-6		
		L13	10		
		D13	6		
		K14	2.097969E-6		
		L14	10		
		D14	6		

Variables

pcct with pumps sol.tkw

Variables

Status	Input	Name	Output	Unit	Comment
		K15	2.097969E-6		
		L15	10		
		D15	6		
		K16	2.097969E-6		
		L16	10		
		D16	6		
		K17	2.097969E-6		
		L17	10		
		D17	6		
		K18	5.174993E-7		
		L18	10		
		D18	8		
		K19	1.747407E-7		
		L19	10		
		D19	10		
		K20	1.747407E-7		
		L20	10		
		D20	10		
		K21	7.196772E-8		
		L21	10		
		D21	12		
		K22	1.799193E-7		
		L22	25		
		D22	12		
		K23	2.62111E-7		
		L23	15		
		D23	10		
		K24	7.762489E-7		
		L24	15		
		D24	8		
		K28	1.799193E-7		
		L28	25		
		D28	12		
		K29	3.598386E-8		
		L29	5		
		D29	12		
		K30	2.587496E-7		

Variables

Status	Input	Name	Output	Unit	Comment
		L30	5		
		D30	8		
		P7	55.8115141		
		P8	55.1741105		
		Q6	5692.26047		
		K	1.85		
		P9	54.1514548		
		Q7	4550.08085		
		P10	53.5507035		
		Q8	3412.99385		
		P11	52.7099948		
		Q9	2275.90685		
		P12	51.7710501		
		Q10	1133.72724		
		P13	35.5926734		
		P14	34.6537286		
		Q17	1133.72724		
		P15	33.81302		
		Q18	2275.90685		
		P16	33.2122686		
		Q19	3412.99385		
		P17	32.1896129		
		Q20	4550.08085		
		P18	31.5522093		
		Q21	5692.26047		
		Q11	1133.72724		
		KCOND	258.2		
		Q12	1142.17961		
		Q13	1137.08700		
		Q14	1137.08700		
		Q15	1142.17961		
		Q16	1133.72724		
		P19	14.3223169		
		Q22	6825.98771	GPM	TOTAL FLOW
		P20	12.8608672		
		Q23	4432.49098		
		P21	11.7117251		

pcct with pumps sol.tkw

Variables

Status	Input	Name	Output	Unit	Comment
		Q24	2164.395544		FLOW TO TOWER 1
		Q25	2393.496725	GPM	K FACTOR FOR COOLING TOWER
		KTWR	632.45		
		Q26	2268.095439	GPM	FLOW TO TOWER 2
		Q27	2164.395544	GPM	FLOW TO TOWER 3
		P6	60.7172774	FT. W.G.	PRESSURE AT NODE 6
		Q5	6825.987715	GPM	TOTAL FLOW
		V5	19.36399980	FT/SEC	FLOW VELOCITY IN PIPE 5, TYPICAL
		V6	16.1478361		
		V7	18.5870803		
		V8	13.9420799		
		V9	14.5266867		
		V10	12.8646549		
		V11	12.8646549		
		V12	12.9605658		
		V13	12.9027789		
		V14	12.9027789		
		V15	12.9605658		
		V16	12.8646549		
		V17	7.23636839		
		V18	14.5266867		
		V19	13.9420799		
		V20	18.5870803		
		V21	16.1478361		
		V22	19.36399980		
		V23	18.1067256		
		V24	13.8149309		
		P4	61.7984297		
		P1	15	FT. W.G.	SUCTION PRESSURE, GIVEN
		Q1	2258.47666		
		P5	60.9695937		
		P2	14.7879125		
		Q2	2287.76079		
		P3	14.366244		
		Q3	2279.75025		
		Q4	4546.23745		
		Q29	4567.51105		

pcct with pumps sol.tkw

Variables

Status	Input	Name	Output	Unit	Comment
		Q30	2279.75025(		
		V1	14.4154330(		
		V4	12.8967916;		
		V29	12.9571407;		
		V30	14.5512184;		

pcct with pumps sol.tkw

FIGURE 9.21
CONDENSATE PUMP

Liberty Pumps®

Mid-Range FL50-Series
1/2 hp

High Head FL60-Series
6/10 hp

Effluent Pumps
1/2" Solids Handling

FL50-Series
48' Max Head

2 YEAR WARRANTY

FL60-Series
63' Max Head

Models:

FL51M	115V, 12a, Manual	
FL51A	115V, 12a, Automatic	CSA
FL52M	208-230V, 7a, Manual	
FL52A	208-230V, 7a, Automatic	
FL61M	115V, 13a, Manual	
FL61A	115V, 13a, Automatic	CSA
FL62M	208-230V, 7a, Manual	
FL62A	208-230V, 7a, Automatic	

Automatic models feature a mercury-free wide-angle float with series plug – allows for manual operation of pump separate from switch.

POWDER COATED TOUGH™

Features:
- Semi-open impeller permits passage of solids without clogging
- Cast iron construction with all stainless and brass fasteners
- 416 stainless steel rotor shaft
- Oil-filled, hermetically sealed motors with thermal overload protection
- Permanently lubricated upper and lower ball bearings
- Unitized, carbon and ceramic shaft seal
- Single float, mercury-free level control with series plug for manual bypass operation – standard on automatic models
- Adjustable pumping range
- 1-1/2" FNPT discharge
- Quick-disconnect 10' standard power cord allows replacement of cord in seconds without breaking seals to motor. (25' length optional)

innovate. evolve.

FL50- & FL60-SERIES
TECHNICAL SPECIFICATIONS

PUMP
The pump(s) shall be model _____ as manufactured by Liberty Pumps, Bergen, NY, or equal.

The pump(s) shall have a capacity of ____ GPM at a total dynamic head of ____ feet.

Motor size shall be ____ horsepower, single phase, 60 hz. and ____ (115 or 208-230) volt operation.

MOTOR
The pump motor shall be of the submersible type, oil filled, hermetically sealed and shall be thermally protected. The overload element shall automatically reset when motor cools. Motor windings shall be of the class B insulation rating.

The rotor shaft shall be made of 416 stainless steel and shall be supported by upper and lower ball bearings.

IMPELLER
The pump shall have a semi-open impeller capable of passing a minimum 1/2" spherical solid and shall be constructed of class 25 or better cast iron.

SEAL
The shaft seal shall be of the carbon/ceramic unitized design, with BUNA N elastomers and stainless housings.

EXTERNAL CONSTRUCTION
The pump volute, legs and motor housing shall be gray iron castings, class 25 or better. All castings shall be epoxy powder coated before assembly. All fasteners shall be of 300-series stainless steel or brass.

LEVEL CONTROL
The pump shall be controlled by an adjustable, mercury-free switch sealed in a polymeric float, and shall have a series plug for manual bypass operation.

DIMENSIONAL DATA:
Weight: FL50/60: 57 LBS.

Height: 13.75"

Major Width: 10.25" (manual models)

Maximum Fluid Temperature: 140° F.

MODELS	HP	VOLTS	PHASE	AMPS	DISCHARGE	AUTOMATIC
FL51M	1/2	115	1	12	1-1/2" FNPT	NO
FL51A	1/2	115	1	12	1-1/2" FNPT	YES
FL52M	1/2	208-230	1	7	1-1/2" FNPT	NO
FL52A	1/2	208-230	1	7	1-1/2" FNPT	YES
FL61M	6/10	115	1	13	1-1/2" FNPT	NO
FL61A	6/10	115	1	13	1-1/2" FNPT	YES
FL62M	6/10	208-230	1	7	1-1/2" FNPT	NO
FL62A	6/10	208-230	1	7	1-1/2" FNPT	YES

10' cord standard on above models. For 25' cord options, add a "-2" suffix to model number. Example: FL61A-2 for Model FL61A with 25' cord.

FL50 PERFORMANCE CURVE
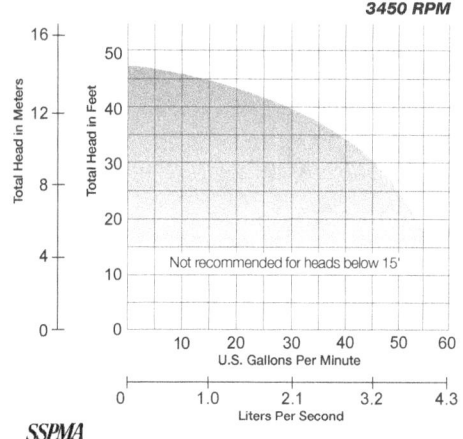

FL60 PERFORMANCE CURVE
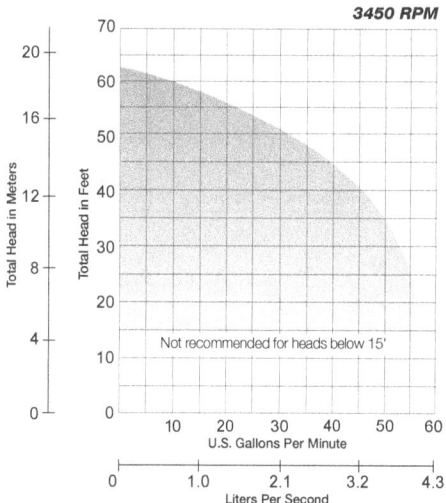

Specifications subject to change without notice.

Liberty Pumps • 7000 Apple Tree Avenue • Bergen, New York 14416 • Phone 800-543-2550 Fax (585) 494-1839
www.libertypumps.com

Pump Specifications

FL60 Series 6/10 hp Submersible Effluent Pump

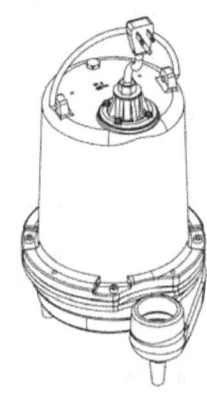

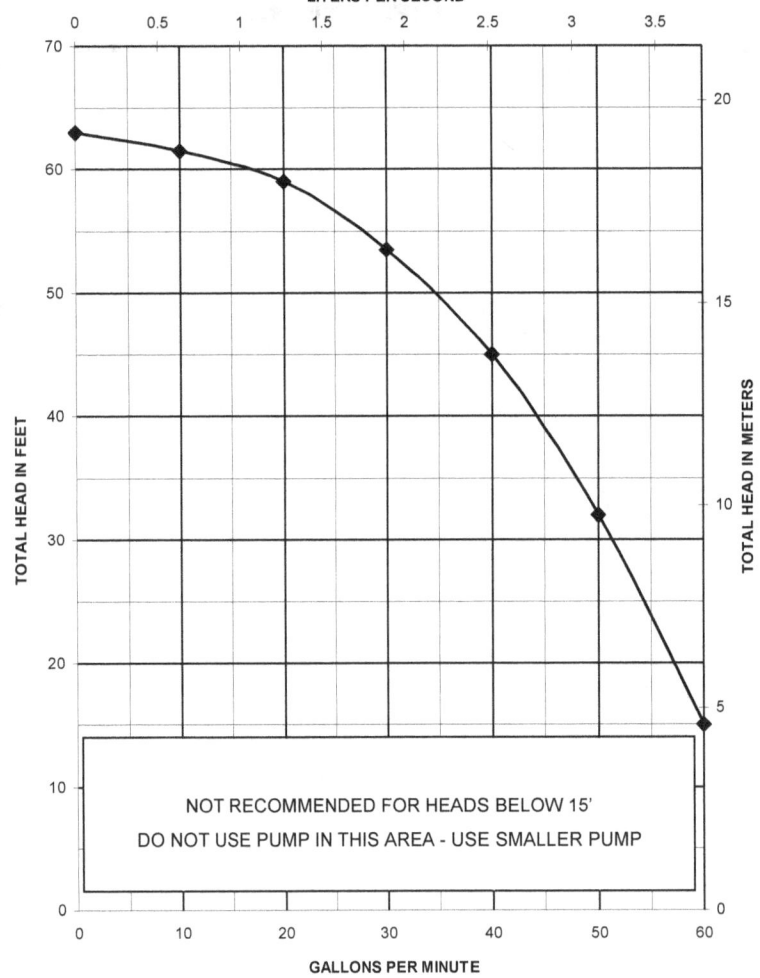

NOT RECOMMENDED FOR HEADS BELOW 15'
DO NOT USE PUMP IN THIS AREA - USE SMALLER PUMP

FL60-Series Dimensional Data

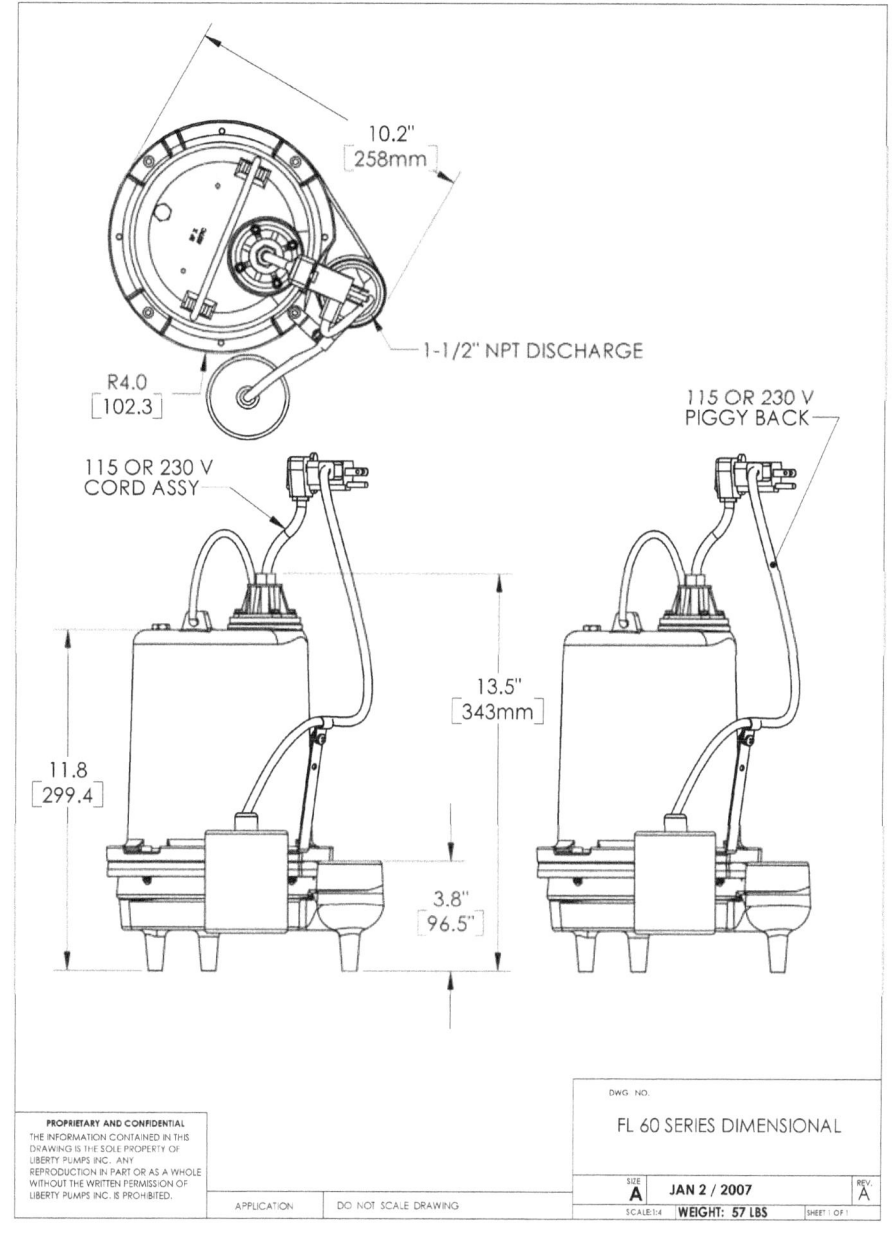

FIGURE 9.20 LIBERTY FL 60 CURVE FIT

Polynomial Regression Worksheet (y = b1 + b2*x + b3*x^2 + ...)

	X	Y	residuals	SUMMARY STATS	
1	0	63	-2.143E-1	order	3
2	10	60	+5.714E-1	N	6
3	20	55.5	-1.429E-1	Syx	1.017700489
4	30	50	-8.571E-1	adj R2	.990928712
5	40	45	+9.286E-1	p	.015474215
6	50	34	-2.857E-1		
7				b1	63.21428571
8					-.411904762
9					.005
10					-.000166667
11					0
12					0
13					0
14					0
15					0

INT TABLE: works   LIBERTY FL60.tkw

9.22 CONDENSATE PUMPS IN PARALLEL

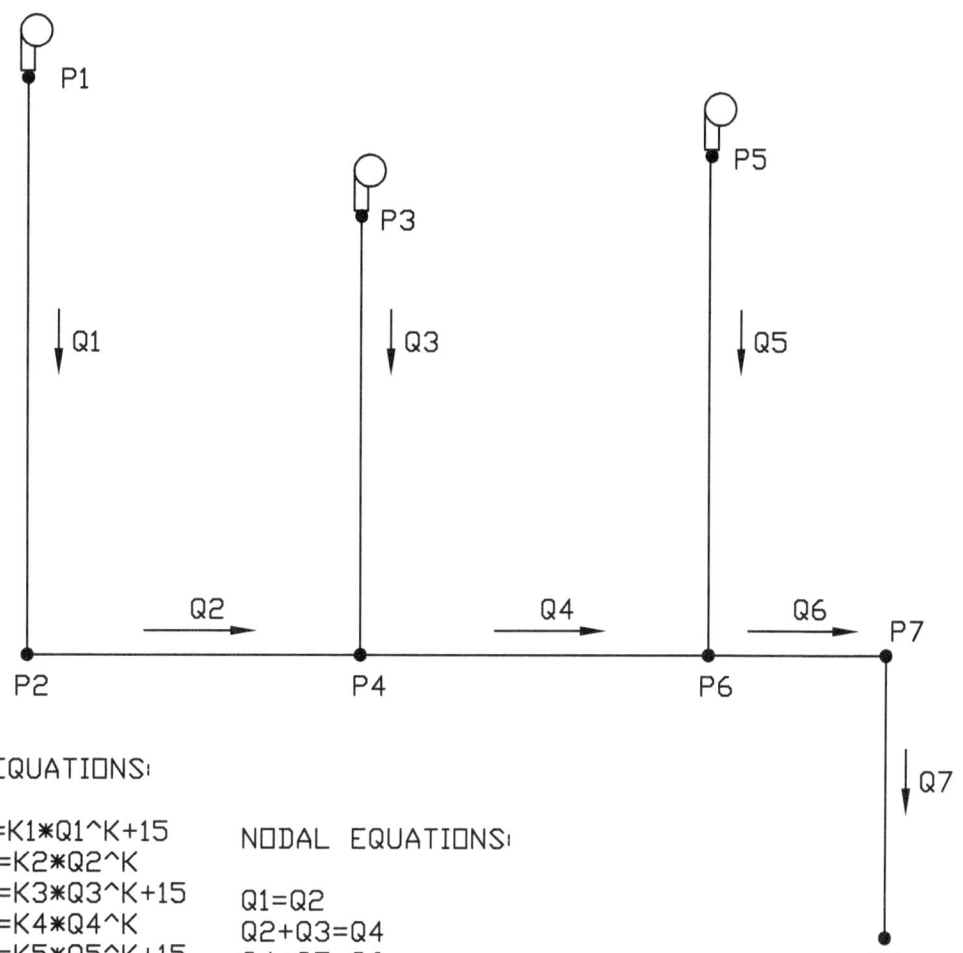

PIPE EQUATIONS:

P1-P2=K1*Q1^K+15
P2-P4=K2*Q2^K
P3-P4=K3*Q3^K+15
P4-P6=K4*Q4^K
P5-P6=K5*Q5^K+15
P6-P7=K6*Q6^K
P7-P8=K7*Q7^K

NODAL EQUATIONS:

Q1=Q2
Q2+Q3=Q4
Q4+Q5=Q6
Q6=Q7

PUMP CURVE:

P1=63.214-0.41190462*Q1+0.005*Q1^2-0.000166667*Q1^3

P3=63.214-0.41190462*Q3+0.005*Q3^2-0.000166667*Q3^3

P5=63.214-0.41190462*Q5+0.005*Q5^2-0.000166667*Q5^3

Equations

```
/*  ESUITE 1.04 */
/*  CONDENSATE PUMPS IN PARALLEL  */
/*  HAZEN WILLIAMS EQUATION FOR FT. OF HEAD  */
K1=0.002083*L1*(100/C)^1.85/(D1^4.8655)
K2=0.002083*L2*(100/C)^1.85/(D2^4.8655)
K3=0.002083*L3*(100/C)^1.85/(D3^4.8655)
K4=0.002083*L4*(100/C)^1.85/(D4^4.8655)
K5=0.002083*L5*(100/C)^1.85/(D5^4.8655)
K6=0.002083*L6*(100/C)^1.85/(D6^4.8655)
K7=0.002083*L7*(100/C)^1.85/(D7^4.8655)
/*  CONSTANTS  */
C=130
K=1.85
/* LINK EQUATIONS */
P1-P2=K1*Q1^K+15
P2-P4=K2*Q2^K
P3-P4=K3*Q3^K+15
P4-P6=K4*Q4^K
P5-P6=K5*Q5^K+15
P6-P7=K6*Q6^K
P7-P8=K7*Q7^K
/* NODAL EQUATIONS */
Q1=Q2
Q2+Q3=Q4
Q4+Q5=Q6
Q6=Q7
/* PUMP CURVE  */
P1=63.214-0.41190462*Q1+0.005*Q1^2-0.000166667^Q1^3
P3=63.214-0.41190462*Q3+0.005*Q3^2-0.000166667^Q3^3
P5=63.214-0.41190462*Q5+0.005*Q5^2-0.000166667^Q5^3
/*  FLOW VELOCITIES  */
V1=0.4085*Q1/(D1^2)
V2=0.4085*Q2/(D2^2)
V3=0.4085*Q3/(D3^2)
V4=0.4085*Q4/(D4^2)
V5=0.4085*Q5/(D5^2)
V6=0.4085*Q6/(D6^2)
V7=0.4085*Q7/(D7^2)
```

```
/*  PIPE LENGTHS AND DIAMETERS  */
L1=34
D1=1
L2=27
D2=1
L3=24
D3=1
L4=30
D4=1
L5=55
D5=1
L6=72
D6=1
L7=38
D7=1
/* GIVEN */
P8=1
```

Results

```
C=130
D1=1
D2=1
D3=1
D4=1
D5=1
D6=1
D7=1
K=1.85
K1=0.0435886
K2=0.0346145
K3=0.0307684
K4=0.0384605
K5=0.070511
K6=0.0923053
K7=0.0487167
L1=34
L2=27
```

L3=24
L4=30
L5=55
L6=72
L7=38
P1=61.2079007
P2=45.2880454
P4=44.557572
P3=60.6337006
P6=40.7254418
P5=59.8902155
P7=14.7233344
P8=1
Q1=5.1983198
Q2=5.1983198
Q3=6.8306849
Q4=12.0290047
Q5=9.0673043
Q6=21.096309
Q7=21.096309
V1=2.1235136
V2=2.1235136
V3=2.7903348
V4=4.9138484
V5=3.7039938
V6=8.6178422
V7=8.6178422

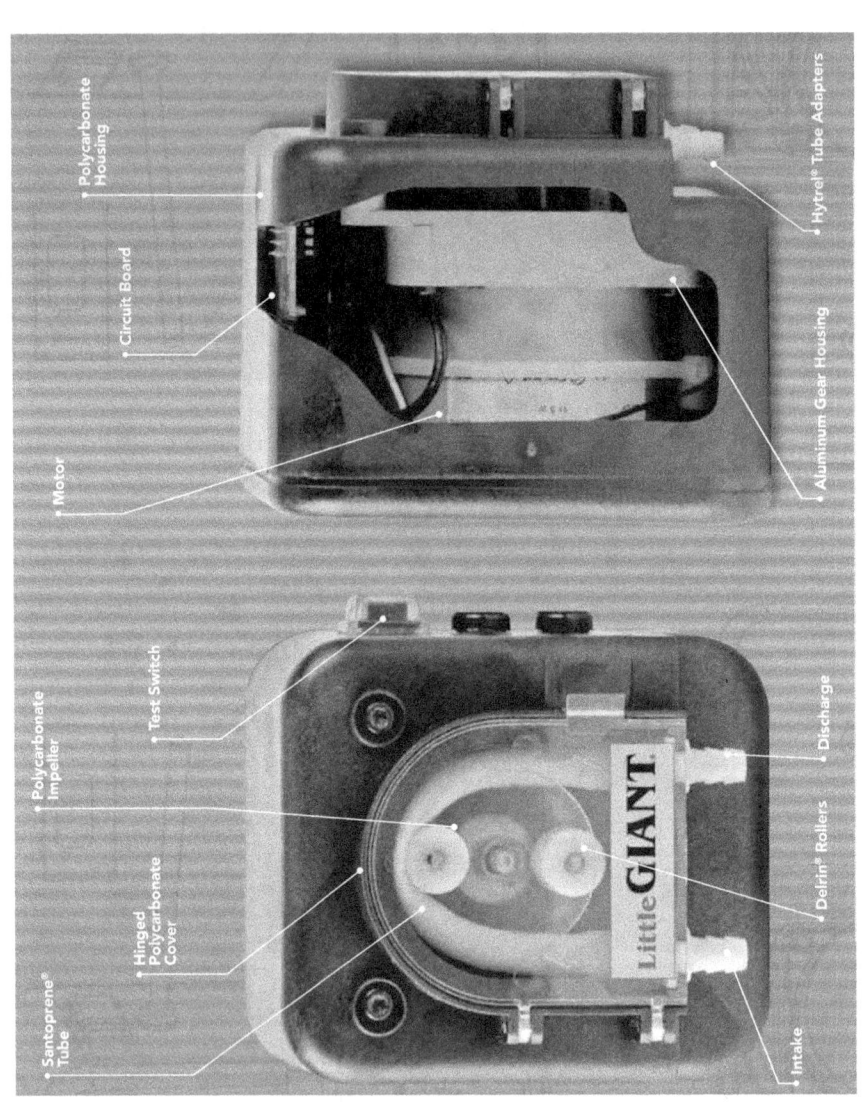

Equations

```
/*  ESUITE 1.04  */
/*  CONDENSATE PUMPS IN PARALLEL  */
/*  HAZEN WILLIAMS EQUATION FOR FT. OF HEAD  */
/*  FIGURE 9.26 PERISTALIC PUMPS GIVEN DISCHARGE PRESSURES  */
K1=0.002083*L1*(100/C)^1.85/(D1^4.8655)
K2=0.002083*L2*(100/C)^1.85/(D2^4.8655)
K3=0.002083*L3*(100/C)^1.85/(D3^4.8655)
K4=0.002083*L4*(100/C)^1.85/(D4^4.8655)
K5=0.002083*L5*(100/C)^1.85/(D5^4.8655)
K6=0.002083*L6*(100/C)^1.85/(D6^4.8655)
K7=0.002083*L7*(100/C)^1.85/(D7^4.8655)
/*  CONSTANTS  */
C=130
K=1.85
/*  LINK EQUATIONS  */
P1-P2=K1*Q1^K+15
P2-P4=K2*Q2^K
P3-P4=K3*Q3^K+15
P4-P6=K4*Q4^K
P5-P6=K5*Q5^K+15
P6-P7=K6*Q6^K
P7-P8=K7*Q7^K
/*  NODAL EQUATIONS  */
Q1=Q2
Q2+Q3=Q4
Q4+Q5=Q6
Q6=Q7
/*  PUMP CURVE  */
P1=46
P3=46
P5=46
/*  FLOW VELOCITIES  */
V1=0.4085*Q1/(D1^2)
V2=0.4085*Q2/(D2^2)
V3=0.4085*Q3/(D3^2)
V4=0.4085*Q4/(D4^2)
V5=0.4085*Q5/(D5^2)
V6=0.4085*Q6/(D6^2)
```

```
V7=0.4085*Q7/(D7^2)
/*  PIPE LENGTHS AND DIAMETERS  */
L1=34
D1=1
L2=27
D2=1
L3=24
D3=1
L4=30
D4=1
L5=55
D5=1
L6=72
D6=1
L7=38
D7=1
/* GIVEN */
P8=0
```

Results

```
C=130
D1=1
D2=1
D3=1
D4=1
D5=1
D6=1
D7=1
K=1.85
K1=0.0435886
K2=0.0346145
K3=0.0307684
K4=0.0384605
K5=0.070511
K6=0.0923053
K7=0.0487167
L1=34
```

```
L2=27
L3=24
L4=30
L5=55
L6=72
L7=38
P2=30.5455484
P4=30.1846603
P6=27.7409396
P7=9.5832337
P1=46
P3=46
P5=46
P8=0
Q1=3.5513992
Q2=3.5513992
Q3=5.8813307
Q4=9.4327299
Q5=7.9417797
Q6=17.3745097
Q7=17.3745097
V1=1.4507466
V2=1.4507466
V3=2.4025236
V4=3.8532702
V5=3.244217
V6=7.0974872
V7=7.0974872
```

Equations

```
/*  ESUITE 1.04 */
/*  CONDENSATE PUMPS IN PARALLEL  */
/*  HAZEN WILLIAMS EQUATION FOR FT. OF HEAD  */
/*  FIGURE 9.27 PERISTALIC PUMPS GIVEN FLOW RATES  */
K1=0.002083*L1*(100/C)^1.85/(D1^4.8655)
K2=0.002083*L2*(100/C)^1.85/(D2^4.8655)
K3=0.002083*L3*(100/C)^1.85/(D3^4.8655)
K4=0.002083*L4*(100/C)^1.85/(D4^4.8655)
K5=0.002083*L5*(100/C)^1.85/(D5^4.8655)
K6=0.002083*L6*(100/C)^1.85/(D6^4.8655)
K7=0.002083*L7*(100/C)^1.85/(D7^4.8655)
/*  CONSTANTS  */
C=130
K=1.85
/* LINK EQUATIONS */
P1-P2=K1*Q1^K+15
P2-P4=K2*Q2^K
P3-P4=K3*Q3^K+15
P4-P6=K4*Q4^K
P5-P6=K5*Q5^K+15
P6-P7=K6*Q6^K
P7-P8=K7*Q7^K
/* NODAL EQUATIONS */
Q1=Q2
Q2+Q3=Q4
Q4+Q5=Q6
Q6=Q7
/* PUMP CURVE  */
Q1=0.0433
Q3=0.0433
Q5=0.0433
/*  FLOW VELOCITIES  */
V1=0.4085*Q1/(D1^2)
V2=0.4085*Q2/(D2^2)
V3=0.4085*Q3/(D3^2)
V4=0.4085*Q4/(D4^2)
V5=0.4085*Q5/(D5^2)
V6=0.4085*Q6/(D6^2)
```

```
V7=0.4085*Q7/(D7^2)
/*  PIPE LENGTHS AND DIAMETERS  */
L1=34
D1=1
L2=27
D2=1
L3=24
D3=1
L4=30
D4=1
L5=55
D5=1
L6=72
D6=1
L7=38
D7=1
/* GIVEN */
P8=0
```

Results

```
C=130
D1=1
D2=1
D3=1
D4=1
D5=1
D6=1
D7=1
K=1.85
K1=0.0435886
K2=0.0346145
K3=0.0307684
K4=0.0384605
K5=0.070511
K6=0.0923053
K7=0.0487167
L1=34
```

L2=27
L3=24
L4=30
L5=55
L6=72
L7=38
P8=0
P7=0.0011165
P6=0.0032319
P4=0.0036483
P2=0.0037522
P1=15.0038831
P3=15.0037406
P5=15.0034437
Q1=0.0433
Q2=0.0433
Q3=0.0433
Q4=0.0866
Q5=0.0433
Q6=0.1299
Q7=0.1299
V1=0.017688
V2=0.017688
V3=0.017688
V4=0.0353761
V5=0.017688
V6=0.0530642
V7=0.0530642

9.28 GRAVITY LINE TO PUMPED LINE

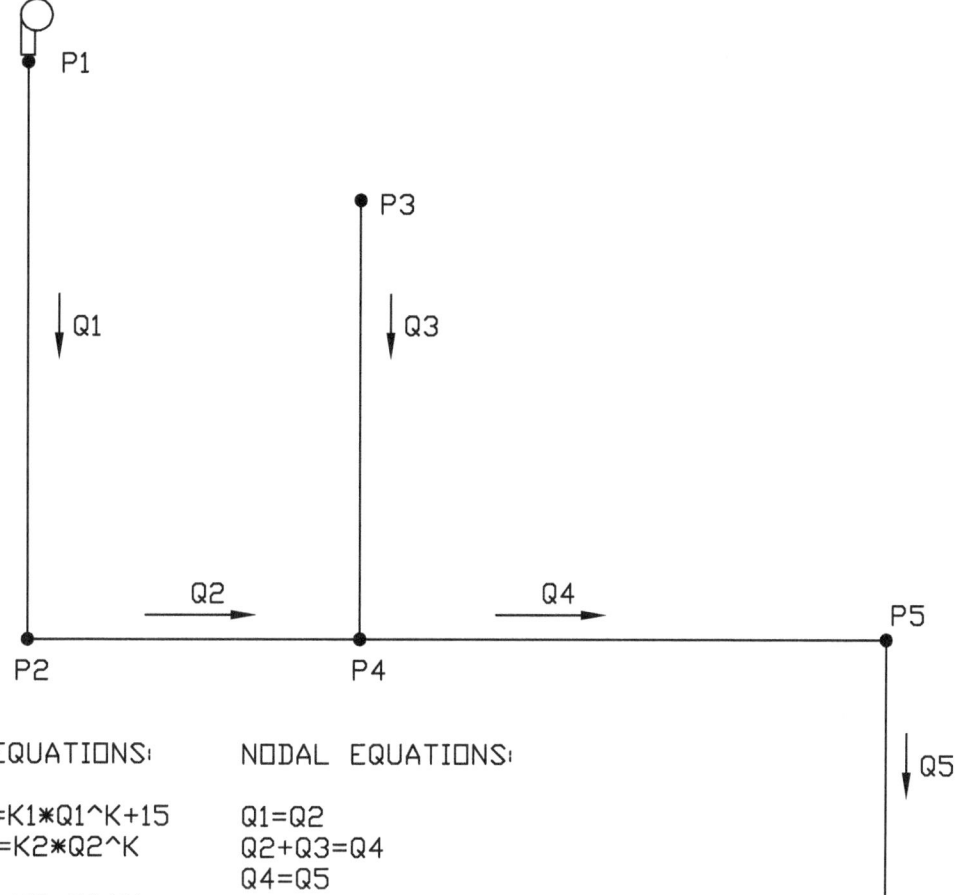

PIPE EQUATIONS: NODAL EQUATIONS:

P1-P2=K1*Q1^K+15   Q1=Q2
P2-P4=K2*Q2^K   Q2+Q3=Q4
P3=5   Q4=Q5
P3-P4=K3*Q3^K
P4-P5=K4*Q4^K
P5-P6=K5*Q5^K
P6=0

PUMP CURVE:

P1=63.214-0.41190462*Q1+0.005*Q1^2-0.000166667*Q1^3

Rule

;>>>/* CONDENSATE PUMPS IN PARALLEL */
;>>>/* HAZEN WILLIAMS EQUATION FOR FT. OF HEAD */
;>>>/* FIG. 9.29 GRAVITY LINE TO PUMPED LINE */
K1=0.002083*L1*(100/C)^1.85/(D1^4.8655)
K2=0.002083*L2*(100/C)^1.85/(D2^4.8655)
K3=0.002083*L3*(100/C)^1.85/(D3^4.8655)
K4=0.002083*L4*(100/C)^1.85/(D4^4.8655)
K5=0.002083*L5*(100/C)^1.85/(D5^4.8655)

;>>>/* CONSTANTS */
C=130
K=2

;>>>/* LINK EQUATIONS */
P1-P2=K1*Q1^K+15
P2-P4=K2*Q2^K
P3=5
P3-P4=K3*Q3^K
P4-P5=K4*Q4^K
P5-P6=K5*Q5^K

;>>>/* NODAL EQUATIONS */
Q1=Q2
Q2+Q3=Q4
Q4=Q5

;>>>/* PUMP CURVE */
P1=63.214-0.41190462*Q1-0.005*Q1^2-0.000166667*Q1^3

;>>>/* FLOW VELOCITIES */
V1=0.4085*Q1/(D1^2)
V2=0.4085*Q2/(D2^2)
V3=0.4085*Q3/(D3^2)
V4=0.4085*Q4/(D4^2)
V5=0.4085*Q5/(D5^2)

;>>>/* PIPE LENGTHS AND DIAMETERS */
L1=34
D1=1
L2=27
D2=1
L3=6
D3=1
L4=30
D4=1
L5=55
D5=1

Rule
;>>>/* GIVEN */
P6=0

Status	Input	Name	Output	Unit	Comment
		K1	.0435886065		
		L1	34		
		C	130		
		D1	1		
		K2	.0346144816		
		L2	27		
		D2	1		
		K3	.007692107		
		L3	6		
		D3	1		
		K4	.0384605356		
		L4	30		
		D4	1		
		K5	.0705109811		
		L5	55		
		D5	1		
		K	2		
		P1	51.2096237	FT. W.G.	PUMP DISCHARGE PRESSURE
		P2	17.8413176	FT. W.G.	PRESSURE IN DRAIN LINE FROM PUMP
		Q1	20.5280672	GPM	PUMP FLOW RATE
		P4	3.25472165	FT. W.G.	PRESSURE AT JUNCTION OF GRAVITY TO PUMPED LINE
		Q2	20.5280672	GPM	PUMP FLOW RATE
		P3	5	FT. W.G.	HEAD AT START OF GRAVITY LINE
		Q3	-15.0629381	GPM	FLOW RATE IN GRAVITY LINE, NOTE DIRECTION.
		P5	2.10599636	FT. W.G.	
		Q4	5.46512915	GPM	TOTAL FLOW RATE
		P6	0	FT. W.G.	FINAL PRESSURE AT END OF LINE
		Q5	5.46512915	GPM	TOTAL FLOW RATE
		V1	8.38571546	FT/SEC	FLOW VELOCITY IN PIPE 1, TYPICAL
		V2	8.38571546		
		V3	-6.15321021		
		V4	2.23250525		
		V5	2.23250525		

FIG. 9.28 GRAVITY TO PUMPED LINE.tkw

CHAPTER 10 – AIR DUCTS

FIG. 10.1 FIND DUCT DIAMETERS

FIG. 10.2 FIND DUCT FLOW RATES

FIG. 10.3 7 EXHAUSTS

FIG. 10.4 HOT EXHAUSTS

Air ducts are used extensively in the HVAC industry. They can be supply, return, exhaust or outside air. Most are built from galvanized steel, and that is what we will focus on. Other materials can be used such as aluminum, stainless steel, and pvc. All those materials just change the friction factor.

For regular air ducts, we will use the equation in the Carrier Duct Design Manual. It is:

DP = 0.03 * f*(L/(d^1.22))*(V/1000)^1.82 = in w.g.

Where f= 0.9 for galvanized steel
L= length of duct, feet
d= duct diameter, inches for round duct, or equivalent round diameter for rectangular duct.
V = air velocity, ft/min.

Note that this is for atmospheric air, such is found in regular HVAC systems. It is NOT suitable for high temperatures and pressures.

Now, V=Q/A, where Q = cfm, A = flow area, sq.ft. A = pi/4*(d/12)^2 and
V = 183.346*Q/(d^2) = fpm

So after some algebra, dp = 1.2317*10^-3*L*(1/d)^1.22*Q^1.82/(d^3.7)

Or, dp = 1.2317*10^-3*L*Q^1.82/(d^4.51) (looks familiar)

Or, dp = KX*Q^k

Where KX=1.2317*10^-3*L*(1/d)^4.51, and k = 1.82

For a duct section 1 with length L1 and diameter d1,
K1= 1.2317*10^-3*L1*(1/d1)^4.51, similar to the way we did it for water pipes.

Just round ducts are considered. Rectangular ducts can of course be analyzed, by finding the equivalent round duct diameter, De.

From SMACNA Duct Design Manual, **De=(4*W*H/pi)^0.5**

Where W = rectangular duct width, H = rectangular duct height, pi = 3.14159. This can be used to provide round duct sizes, or, if a round duct size is calculated, a rectangular size can be determined, if one dimension is known. Such as if De calculated =27.3" and maximum height = 12", then W = pi*(De)^2/(4*H) = 48.78 → use 49" wide.

FIRST PROBLEM: 10.1 , Solve for Duct Diameters

Suppose we have an air distribution system with a fan delivering 0.6" w.g. pressure. The outlet CFMs are known, but the sizes are not. Also, we want to limit the friction loss to a certain rate, say 0.10"/100'. So, from say point P6 to P8 with a duct length of L6, the pressure drop is = 0.1"/100'. Or, (P6-P8)*100/L6=0.10

See Diagram 10.01, Rules and Sol sheets. Ducts d1, d2, d3, d4, d5, d6, d7, d8, d9, d10 and d11 are all sized. Also, all are sized to meet the maximum friction rate specified. If you want to be sure, check out P6 and P8. (0.544-0.529)*100/15 = 0.10, just as promised. The same technique can be used to size the ducts for a set velocity. It is either or, not both criteria.

FIND DUCT DIAMETERS

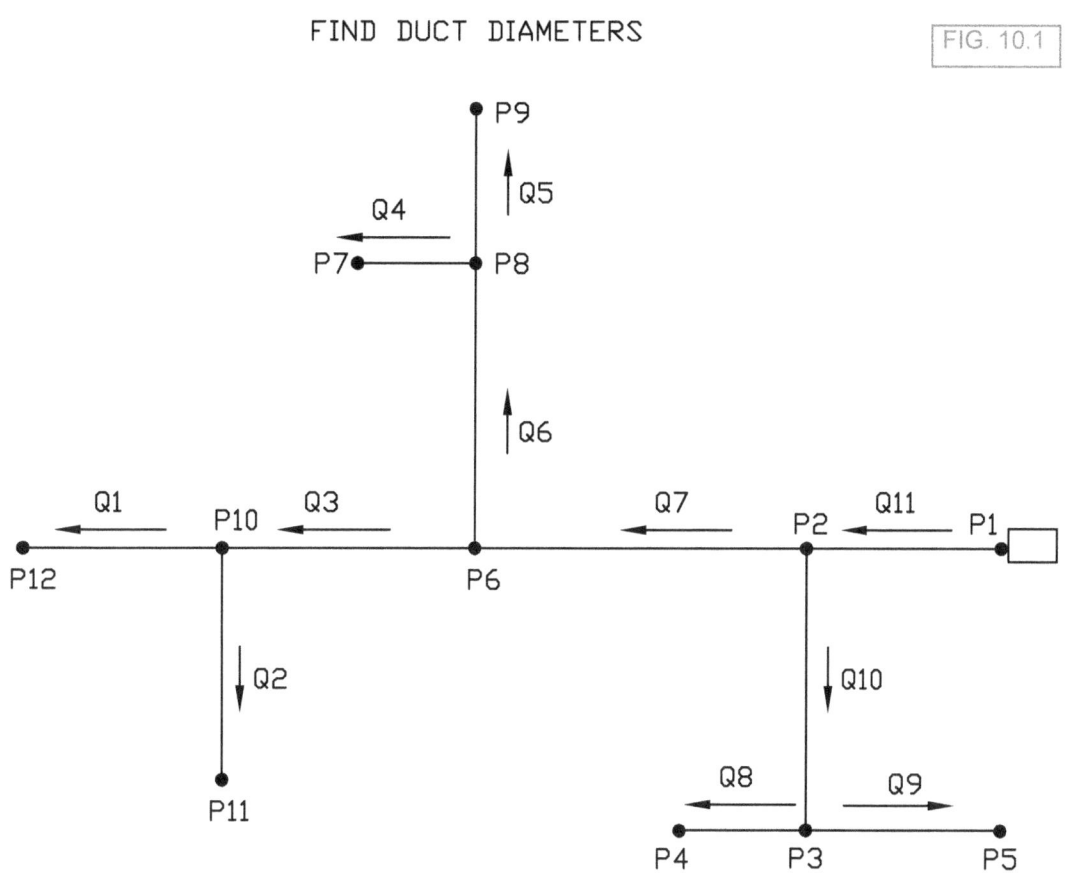

FIG. 10.1

DUCT LINKS:

P1-P2=K11*Q11^K
P2-P3=K10*Q10^K
P3-P4=K8*Q8^K
P3-P5=K9*Q9^K
P2-P6=K7*Q7^K
P6-P8=K6*Q6^K
P8-P7=K4*Q4^K
P8-P9=K5*Q5^K
P6-P10=K3*Q3^K
P10-P11=K2*Q2^K
P10-P12=K1*Q1^K

NODAL EQUATIONS:

Q11=Q10+Q7
Q10=Q8+Q9
Q7=Q3+Q6
Q6=Q4+Q5
Q3=Q1+Q2

GIVEN FLOWS:

Q1=350
Q2=450
Q4=450
Q5=300
Q8=175
Q9=250

PD RATE:

(P1-P2)*100/L11=0.10
(P2-P3)*100/L10=0.10
(P3-P4)*100/L8=0.10
(P3-P5)*100/L9=0.10
(P2-P6)*100/L7=0.10
(P6-P8)*100/L6=0.10
(P8-P7)*100/L4=0.10
(P8-P9)*100/L5=0.10
(P6-P10)*100/L3=0.10
(P10-P11)*100/L2=0.10
(P10-P12)*100/L1=0.10

Status   Rule

Satisfi K1=1.23174*10^-3*L1*(1/d1)^4.51
Satisfi K2=1.23174*10^-3*L2*(1/d2)^4.51
Satisfi K3=1.23174*10^-3*L3*(1/d3)^4.51
Satisfi K4=1.23174*10^-3*L4*(1/d4)^4.51
Satisfi K5=1.23174*10^-3*L5*(1/d5)^4.51
Satisfi K6=1.23174*10^-3*L6*(1/d6)^4.51
Satisfi K7=1.23174*10^-3*L7*(1/d7)^4.51
Satisfi K8=1.23174*10^-3*L8*(1/d8)^4.51
Satisfi K9=1.23174*10^-3*L9*(1/d9)^4.51
Satisfi K10=1.23174*10^-3*L10*(1/d10)^4.51
Satisfi K11=1.23174*10^-3*L11*(1/d11)^4.51

Comm ;>>>/* LINK EQUATIONS */
Satisfi P1-P2=K11*Q11^K
Satisfi P2-P3=K10*Q10^K
Satisfi P3-P5=K9*Q9^K
Satisfi P3-P4=K8*Q8^K
Satisfi P2-P6=K7*Q7^K
Satisfi P6-P8=K6*Q6^K
Satisfi P8-P7=K4*Q4^K
Satisfi P8-P9=K5*Q5^K
Satisfi P6-P10=K3*Q3^K
Satisfi P10-P11=K2*Q2^K
Satisfi P10-P12=K1*Q1^K

Comm ;>>>/* NODAL EQUATIONS */
Satisfi Q11=Q7+Q10
Satisfi Q10=Q8+Q9
Satisfi Q7=Q6+Q3
Satisfi Q6=Q4+Q5
Satisfi Q3=Q1+Q2

Comm ;>>>/* AIR VELOCITIES */
Satisfi V1=183.346*Q1/(d1^2)
Satisfi V2=183.346*Q2/(d2^2)
Satisfi V3=183.346*Q3/(d3^2)
Satisfi V4=183.346*Q4/(d4^2)

Status   Rule
Satisfi V5=183.346*Q5/(d5^2)
Satisfi V6=183.346*Q6/(d6^2)
Satisfi V7=183.346*Q7/(d7^2)
Satisfi V8=183.346*Q8/(d8^2)
Satisfi V9=183.346*Q9/(d9^2)
Satisfi V10=183.346*Q10/(d10^2)
Satisfi V11=183.346*Q11/(d11^2)

Comm ;>>/* MAX PRESS DROP */
Satisfi (P1-P2)*100/L11=0.1
Satisfi (P2-P3)*100/L10=0.1
Satisfi (P3-P5)*100/L9=0.1
Satisfi (P3-P4)*100/L8=0.1
Satisfi (P2-P6)*100/L7=0.1
Satisfi (P6-P8)*100/L6=0.1
Satisfi (P8-P7)*100/L4=0.1
Satisfi (P8-P9)*100/L5=0.1
Satisfi (P6-P10)*100/L3=0.1
Satisfi (P10-P11)*100/L2=0.1
Satisfi (P10-P12)*100/L1=0.1

Status	Input	Name	Output	Unit	Comment
	20	K1	4.686258E-	FEET	LENGTH OF DUCT 1
		L1	11.1358527	INCHES	DIAMETER OF DUCT 1
		d1	7.8067E-8		
	15	K3			
		L3	15.5455448	INCHES	
		d3	8.898247E-8		
	6	K4			
		L4	12.3244763	INCHES	
		d4	2.481586E-		
	8	K5			
		L5	10.4642295		
		d5	8.779702E-8		
	15	K6			
		L6	15.1458977		
		d6	2.342543E-8		
	15	K7			
		L7	20.3011981		
		d7	8.273147E-		
	10	K8			
		L8	8.41867073		
		d8	1.729056E-		
	4	K9			
		L9	9.72196043		
		d9	5.101452E-		
	31	K10			
		L10	12.0434517		
		d10	4.119569E-8		
	41	K11			
		L11	22.3866532		
	.6	d11			
		P1	.559	IN. W.G.	SUPPLY PRESSURE AT FAN
		P2	1975	IN. W.G.	
	1.82	Q11		CFM	TOTAL AIR FLOW
		K			
		P3	.528	IN. W.G.	
		Q10	425	CFM	
		P5	.524	IN. W.G.	PRESSURE AT OUTLET

Variables   air ducts find d.tkw

Status	Input	Name	Output	Unit	Comment
	250	Q9		CFM	AIR FLOW, GIVEN
		P4	.518	IN. W.G.	PRESSURE AT OUTLET
	175	Q8		CFM	AIR FLOW, GIVEN
		P6	.544		
		Q7	1550		
		P8	.529		
		Q6	750	CFM	
		P7	.523	IN. W.G.	PRESSURE AT OUTLET
	450	Q4		CFM	AIR FLOW, GIVEN
		P9	.521	IN. W.G.	PRESSURE AT OUTLET
	300	Q5		CFM	AIR FLOW, GIVEN
		P10	.529		
		Q3	800	CFM	
		P11	.499	IN. W.G.	PRESSURE AT OUTLET
		K2	4.449124E-?		GIVEN AIR FLOW
	450	Q2		CFM	GIVEN AIR FLOW
		P12	.509	IN. W.G.	PRESSURE AT OUTLET
	350	Q1		CFM	AIR FLOW AT END OF LINE, GIVEN
	30	L2			
		d2	12.3244763	INCHES	AIR VELOCITY IN DUCT 1
		V1	517.478758	FT/MIN	
		V2	543.184044	FT/MIN	
		V3	606.945271	FT/MIN	
		V4	543.184044	FT/MIN	
		V5	502.317351	FT/MIN	
		V6	599.435778	FT/MIN	
		V7	689.540526	FT/MIN	
		V8	452.712451	FT/MIN	
		V9	484.957577	FT/MIN	
		V10	537.227727	FT/MIN	
		V11	722.537172	FT/MIN	

Variables

air ducts find d.tkw

SECOND PROBLEM: 10.2, Solve for exhaust air flows.

This problem is for 4 air exhausts, with approximately the same flow rate. That is the goal. The 1st flow rate is given, find the other 3, with duct sizes and lengths given. An additional wrinkle added is losses to entrys. For an open duct, the entry loss for air = the loss due to accelerating the air = the velocity pressure = $V_p = (v/4005)^2$, where v=ft/min. Also, for a 30 Deg. Angled entry into a duct, the loss = $0.18*V_p$. This data is from Industrial Ventilation. Total entry loss for a typical inlet = $V_p + 0.18*V_p = 1.18*V_p$.

See diagram 10.2, Rules and Sol sheets. Note that the main sizes must be at least 3" larger in diameter than the branches for the air flow quantities to be close. The main and branch sizes can be adjusted to give best performance, while keeping duct sizes reasonable.

Also note that the inlet pressures are all set to = 0, atmospheric pressure.

4 EXHAUSTS

FIG. 10.2

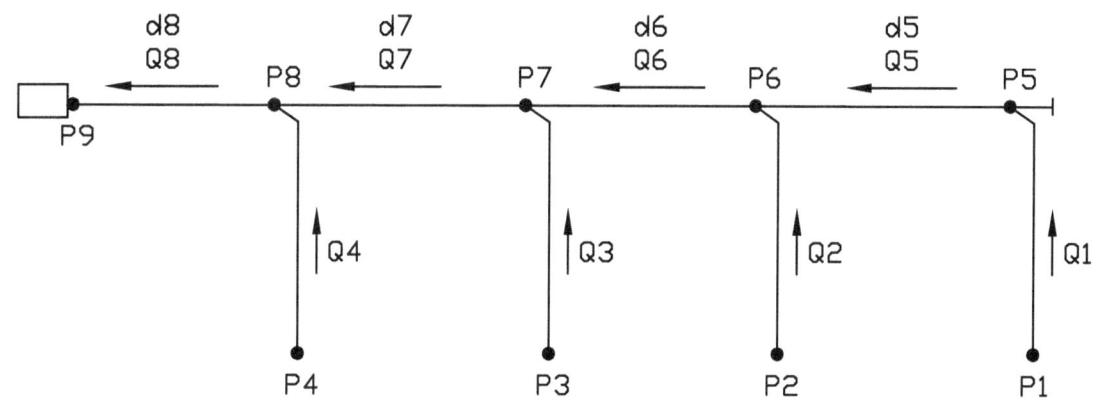

DUCT LINKS:

P1-P5=K1*Q1^K+1.18*VP1
P5-P6=K5*Q5^K
P2-P6=K2*Q2^K+1.18*VP2
P6-P7=K6*Q6^K
P3-P7=K3*Q3^K+1.18*VP3
P7-P8=K7*Q7^K
P4-P8=K4*Q4^K+1.18*VP4
P8-P9=K8*Q8^K

NODAL EQUATIONS:

Q1=Q5
Q5+Q2=Q6
Q6+Q3=Q7
Q7+Q4=Q8

GIVEN FLOWS:

Q1=300

Rule

;>>>/* 08/19/14/* DUCT LENGTH CONSTANTS */14 */
;>>>/* 4 EXHAUST DUCTS */
K1=1.23174*10^-3*L1*(1/d1)^4.51
K2=1.23174*10^-3*L2*(1/d2)^4.51
K3=1.23174*10^-3*L3*(1/d3)^4.51
K4=1.23174*10^-3*L4*(1/d4)^4.51
K5=1.23174*10^-3*L5*(1/d5)^4.51
K6=1.23174*10^-3*L6*(1/d6)^4.51
K8=1.23174*10^-3*L8*(1/d8)^4.51
K7=1.2317*10^-3*L7*(1/d7)^4.51
;>>>/* AREAS */
A1=0.785*(d1/12)^2
A2=0.785*(d2/12)^2
A3=0.785*(d3/12)^2
A4=0.785*(d4/12)^2
A5=0.785*(d5/12)^2
A6=0.785*(d6/12)^2
A7=0.785*(d7/12)^2
A8=0.785*(d8/12)^2

;>>>/* VELOCITIES */
V1=Q1/A1
V2=Q2/A2
V3=Q3/A3
V4=Q4/A4
V5=Q5/A5
V6=Q6/A6
V7=Q7/A7
V8=Q8/A8

;>>>/* VELOCITY PRESSURES */
VP1=(V1/4005)^2
VP2=(V2/4005)^2
VP3=(V3/4005)^2
VP4=(V4/4005)^2
VP5=(V5/4005)^2
VP6=(V6/4005)^2
VP7=(V7/4005)^2
VP8=(V8/4005)^2

;>>>/* LINKS */
P1-P5=K1*Q1^K+1.18*VP1
P5-P6=K5*Q5^K
P2-P6=K2*Q2^K+1.18*VP2
P6-P7=K6*Q6^K
P3-P7=K3*Q3^K+1.18*VP3
P7-P8=K7*Q7^K
P4-P8=K4*Q4^K+1.18*VP4

Rule
P8-P9=K8*Q8^K

;>>>/* DUCT LENGTHS */
L1=15
L2=15
L3=15
L4=15
L5=20
L6=20
L7=20
L8=76

;>>>/* NODAL EQUATIONS */
Q1=Q5
Q5+Q2=Q6
Q6+Q3=Q7
Q7+Q4=Q8

Status	Input	Name	Output	Unit	Comment
		K1	3.558932E-5		
	4	L1	15		
		d1		inches	diameter of 1st pickup pipe
		K2	3.558803E-5		
	4	L2	15		
		d2		inches	diameter of 2nd pickup pipe
		K3	3.558825E-5		
	4	L3	15		
		d3		inches	diameter of 3rd pickup pipe
		K4	3.558942E-5		
	4	L4	15		
		d4		inches	diameter of 4th pickup pipe
		K5	3.803268E-6		
	7	L5	20		
		d5		inches	diameter at main
		K6	4.952952E-7		
	11	L6	20		
		d6		inches	diameter at main
		K8	6.342957E-7		
	14	L8	76		
		d8		inches	diameter at main
		A1	.0872222223	sq. ft.	flow area of duct 1, typical
		A2	.0872222223		
		A3	.0872222223		
		A4	.0872222223		
		A5	.2671180556		
		A6	.6596180556		
		A7	.785		
	12	d7		inches	diameter at main
		A8	1.06847222		
		V1	3439.49044	ft./min	air velocity
	300	Q1		cfm	air flow at 1st pickup pipe, given
		V2	3548.12841		
		Q2	309.475645	cfm	air flow at 2nd pickup pipe
		V3	3598.47053		
		Q3	313.866596	cfm	air flow at 3rd pickup pipe
		V4	3669.76410		

Variables
4 exhausts SOL.tkw

Status	Input	Name	Output	Unit	Comment
		Q4	320.0849799	cfm	air flow at 4th pickup pipe
		V5	1123.098921		
		Q5	300		
		V6	923.9826590		
		Q6	609.4756458		
		V7	1176.232158		
		Q7	923.3422411		
		V8	1163.743145		
		Q8	1243.427222	cfm	total air flow
		VP1	.7375359158	in. w.g.	velocity pressure in pipe 1, typical
		VP2	.7848625714		
		VP3	.8072923933		
		VP4	.8395977288		
		VP5	.078637732		
		VP6	.053225849		
		VP7	.086254359		
		VP8	.084432418		
	0	P1			
		P5	-2.01760145		
	1.82	K			
		P6	-2.1402091		
	0	P2			
		P7	-2.19821667		
	0	P3			
		P8	-2.28165831		
		K7	3.345205E-7		
	0	P4			
		P9	-2.55361569	in. w.g.	pressure at fan
		L7	20		

Variables

4 exhausts SOL.tkw

THIRD PROBLEM: 10.3 More exhausts

This is for 7 exhausts that do not all directly connect to a main. It could represent an existing duct system that we want to check for performance. To solve for all the flows, either 1 flow can be provided, or the suction pressure at the fan. Note that some inlets have just the Vp loss, while others have 1.18*Vp loss. The suction pressure of the fan can be changed to see how the perfomance changes.

7 EXHAUSTS

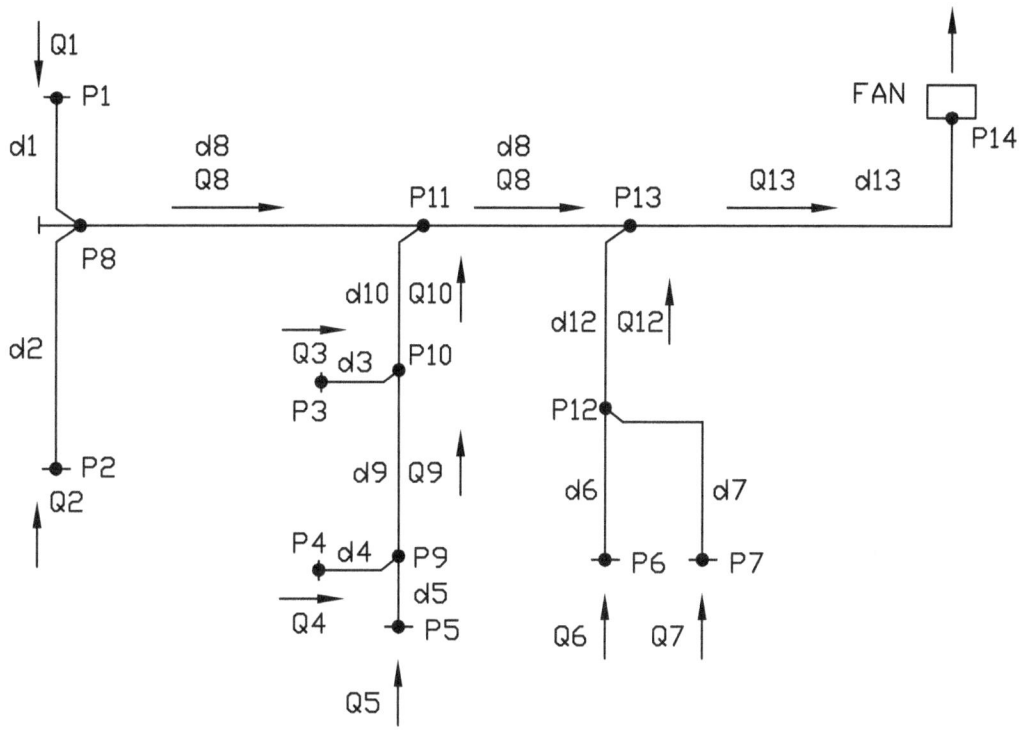

FIG. 10.3

DUCT LINKS:

P1-P8=K1*Q1^K+1.18*VP1
P2-P8=K2*Q2^K+1.18*VP2
P3-P10=K3*Q3^K+1.18*VP3
P4-P9=K4*Q4^K+1.18*VP4
P8-P11=K8*Q8^K
P5-P9=K5*Q5^K+VP5
P4-P9=K4*Q4^K+1.18*VP4
P10-P11=K10*Q19^K+0.18*VP10
P11-P13=K11*Q11^K
P6-P12=K6*Q6^K+VP6
P7-P12=K7*Q7^K+1.18*VP7
P12-P13=K12*Q12^K+0.18*VP12
P13-P14=K13*Q13^K

NODAL EQUATIONS:

Q1+Q2=Q8
Q4+Q5=Q9
Q3+Q9=Q10
Q8+Q10=Q11
Q6+Q7=Q12
Q12+Q11=Q13

GIVEN PRESSURES:

P1=0   P5=0
P2=0   P6=0
P3=0   P7=0
P4=0   P14=-2.0

10-15

Rule

;>>>/* DUCT LENGTH CONSTANTS 8/20/14 */

$K1 = 1.23174 \times 10^{-3} \times L1 \times (1/d1)^{4.51}$
$K2 = 1.23174 \times 10^{-3} \times L2 \times (1/d2)^{4.51}$
$K3 = 1.23174 \times 10^{-3} \times L3 \times (1/d3)^{4.51}$
$K4 = 1.23174 \times 10^{-3} \times L4 \times (1/d4)^{4.51}$
$K5 = 1.23174 \times 10^{-3} \times L5 \times (1/d5)^{4.51}$
$K6 = 1.23174 \times 10^{-3} \times L6 \times (1/d6)^{4.51}$
$K7 = 1.23174 \times 10^{-3} \times L7 \times (1/d7)^{4.51}$
$K8 = 1.23174 \times 10^{-3} \times L8 \times (1/d8)^{4.51}$
$K9 = 1.23174 \times 10^{-3} \times L9 \times (1/d9)^{4.51}$
$K10 = 1.23174 \times 10^{-3} \times L10 \times (1/d10)^{4.51}$
$K11 = 1.23174 \times 10^{-3} \times L11 \times (1/d11)^{4.51}$
$K12 = 1.23174 \times 10^{-3} \times L12 \times (1/d12)^{4.51}$
$K13 = 1.23174 \times 10^{-3} \times L13 \times (1/d13)^{4.51}$

;>>>/* AREAS */
$A1 = 0.785 \times (d1/12)^2$
$A2 = 0.785 \times (d2/12)^2$
$A3 = 0.785 \times (d3/12)^2$
$A4 = 0.785 \times (d4/12)^2$
$A5 = 0.785 \times (d5/12)^2$
$A6 = 0.785 \times (d6/12)^2$
$A7 = 0.785 \times (d7/12)^2$
$A8 = 0.785 \times (d8/12)^2$
$A9 = 0.785 \times (d9/12)^2$
$A10 = 0.785 \times (d10/12)^2$
$A11 = 0.785 \times (d11/12)^2$
$A12 = 0.785 \times (d12/12)^2$
$A13 = 0.785 \times (d13/12)^2$

;>>>/* VELOCITIES */
V1=Q1/A1
V2=Q2/A2
V3=Q3/A3
V4=Q4/A4
V5=Q5/A5
V6=Q6/A6
V7=Q7/A7
V8=Q8/A8
V9=Q9/A9
V10=Q10/A10
V11=Q11/A11
V12=Q12/A12
V13=Q13/A13

;>>>/* VELOCITY PRESSURES */
$VP1 = (V1/4005)^2$
$VP2 = (V2/4005)^2$

Rule
VP3=(V3/4005)^2
VP4=(V4/4005)^2
VP5=(V5/4005)^2
VP6=(V6/4005)^2
VP7=(V7/4005)^2
VP8=(V8/4005)^2
VP9=(V9/4005)^2
VP10=(V10/4005)^2
VP11=(V11/4005)^2
VP12=(V12/4005)^2
VP13=(V13/4005)^2

;>>>/* LINKS */
P1-P8=K1*Q1^K+1.18*VP1
P2-P8=K2*Q2^K+1.18*VP2
P8-P11=K8*Q8^K
P5-P9=K5*Q5^K+VP5
P4-P9=K4*Q4^K+1.18*VP4
P9-P10=K9*Q9^K
P3-P10=K3*Q3^K+1.18*VP3
P10-P11=K10*Q10^K+0.18*VP10
P11-P13=K11*Q11^K
P6-P12=K6*Q6^K+VP6
P7-P12=K7*Q7^K+1.18*VP7
P12-P13=K12*Q12^K+.18*VP12
P13-P14=K13*Q13^K

;>>>/* NODAL EQUATIONS */
Q1+Q2=Q8
Q4+Q5=Q9
Q3+Q9=Q10
Q8+Q10=Q11
Q6+Q7=Q12
Q11+Q12=Q13

Variables

Status	Input	Name	Output	Unit	Comment
	15	K1	5.716736E-6		
	6	L1		FEET	LENGTH OF DUCT 1, TYPICAL
		d1	6.860084E-6	inches	DIAMETER OF DUCT 1, TYPICAL
	18	K2			
	6	L2			
		d2	2.662287E-6	inches	
	14	K3			
	7	L3			
		d3	1.145449E-6	inches	
	11	K4			
	8	L4			
		d4	7.606535E-7	inches	
	4	K5			
	7	L5			
		d5	4.705347E-7	inches	
	19	K6			
	11	L6			
		d6	1.752673E-7	inches	
	21	K8			
	14	L8			
		d8	2.19625	inches	
		A1	.19625	SQ.FT.	FLOW AREA OF DUCT 1, TYPICAL
		A2	.267118055		
		A3	.348888888		
		A4	.267118055		
		A5	.659618055		
		A6	.785		
		A7		inches	
	12	A8	1.06847222		
		V1	22162.7694	FPM	AIR VELOCITY
		Q1	424.443493	cfm	
		V2	2056.67574	FPM	AIR VELOCITY
		Q2	403.622614	cfm	
		V3	2025.99781	FPM	
		Q3	541.180595	cfm	
		V4	1768.98543	FPM	

7 exhausts.tkw

Status	Input	Name	Output	Unit	Comment
		Q4	617.179362	cfm	
		V5	2148.06536	FPM	
		Q5	573.787043		
		V6	1827.90993	FPM	
		Q6	1205.72239		
		V7	1754.10661	FPM	
		Q7	1376.97369		
		V8	775.000127	FPM	
		Q8	828.066108	cfm	
		VP1	.291618713	in. w.g.	VELOCITY PRESSURE
		VP2	.263710006		
		VP3	.255901541		
		VP4	.195093802		
		VP5	.287666933		
		VP6	.208307327		
		VP7	.191825761		
		VP8	.037445402		
	0	P1			
	0	P5			
	1.82	K			
	0	P6			
	0	P2			
	0	P7			
	0	P3			
		P8	-.690674481	IN. W.G.	
		K7	3.34535E-7		
	0	P4			
		P9	-.367467284	in. w.g.	
	20	L7			
		K9	4.683491E-7		
	28	L9			
	12	d9			
		K10	1.836134E-7		
	22	L10			
	14	d10			
		K11	2.448239E-7		
	21	L11			

Variables

7 exhausts.tkw

Status	Input	Name	Output	Unit	Comment
	13	d11			
		K12	3.679886E-		
	22	L12			
	12	d12			
		K13	1.553878E-		
	34	L13			
	16	d13			
		A9	.785		
		A10	1.06847222		
		A11	.921284722		
		A12	.785		
		A13	1.39555555		
		V9	1517.15465		
		Q9	1190.96640		
		V10	1621.14369		
		Q10	1732.14700		
		V11	2778.95969		
		Q11	2560.21310		
		V12	3290.05871		
		Q12	2582.69609		
		V13	3685.20563		
		Q13	5142.9092	CFM	TOTAL AIR FLOW
		VP9	.143500914		
		VP10	.163846806		
		VP11	.481459161		
		VP12	.674842236		
		VP13	.846678267		
		P11	-.726532042		
		P10	-.553119123		
		P13	-1.111729531		
		P12	-.399053452		
	-2	P14		IN. W.G.	SUCTION PRESSURE AT FAN, GIVEN

Variables

7 exhausts.tkw

FOURTH PROBLEM: 10.4 **Hot exhaust**

This problem takes into account air that is not at normal comfort conditions, such as for an industrial oven exhaust system. We will use Darcy's equation:
DP=f*(L/D)*V^2/(2*g), where D is in feet, and for air flow velocity in ft/sec, g = 32.2 ft/sec^2 and Q = ft^3/sec. We want this equation in terms of cfm and inches of duct diameter.

Now, D=d/12, where d is diameter in inches.

Cfm = Vel. x area, with velocity in ft/min and area in sq.ft. Then, Vel.= cfm/area.

Area in sq.ft. = (pi/4)*(d/12)^2. Then, Vel (ft/min) = cfm/((pi/4)*(d/12)^2) = 183.346*cfm/(d^2).

For velocity in ft/sec, Vel = (183.346/60)*cfm/(d^2) = 3.0557*cfm/(d^2).

Now V^2 = 9.33776*cfm^2/(d^4). Then, V^2/(2xg) = 0.14499*cfm^2/(d^4).

Then DP = f*L*(12/d)*0.14499*cfm^2/(d^4) = 1.7399*f*L*cfm^2/(d^5), or
DP = 1.7399*f*L*Q^2*(1/d)^5, where Q = cfm.

The friction factor f, is calculated from the Shacham equation. To do this, we must have the viscosity and density of the fluid (air).

From Crane Technical Paper 410, **Re=123.9*d*V*rho/mu**, where
d=internal diameter of pipe (or duct), inches
V = mean velocity of flow, feet/sec.
Rho = weight density of fluid, lb/cu. Ft.
mu = absolute (dynamic) viscosity of fluid, in centipoises.

Also from Crane, viscosity of air is computed from **Sutherland's formula** as follows:

Mu = muo*(0.555*To+C)/(0.555*T+C)*(T/To)^1.5 = viscosity in centipoises, from 0 to 1000 Deg.F.
Muo = viscosity in centipoises at temperature To.
T = absolute temperature, Deg.R for which viscosity is desired.
To = Absolute temperature, Deg.R for which viscosity is known.
C = Sutherland's constant.

We will choose for our base temperature, To = 100 Deg.F. (This is from the graph)
mu for air at 100 Deg.F = 0.016+0.75*(0.02-0.016)=0.019 centipoise. Then T=100+460 = 560 Deg.R and C = 120.
Then mu = 0.019*((0.555*560+120)/(.555*(T+460)+120)*((T+460)/560)^1.5
Where t = temperature of air, Deg.F.

Then **mu = 0.019*(430.8/(.555*(t+460)+120)*((t+460)/560)^1.5**

This is input into the air duct equations to compute air viscosity.

For air density: From Piping Handbook by Sabin Crocker, 1945,
Density of air = (2.6983*P)/(459.2+t) lb/cu. Ft. = rho
P = absolute pressure in psi = 14.7 psi (at sea level).
t = air temperature, Deg.F, from 0 to 3000 Deg.F.

Then, density (rho) = (2.6983*14.7)/(459.2+t) = **39.665/(459.2+t) = lb/cu.ft.**

Now, DP is in terms of ft. of fluid, which in this case is air. It must be adjusted to obtain " water gage. To convert it, we must multiply DP by the ratio of the fluid density (air) to the density of water. We know the density (rho from above), and the density of water = 62.4 lb/cu.ft.. Now ft. of air = 12" of air, then

DP = 1.7399*12*f*L*Q^2*(1/d)^5*rho/62.4

Or DP = 0.3346*f*L*Q^2*(1/d)^5*rho

Let K = 0.3346*f*L*(1/d)^5*rho, then DP = K*Q^2. This is our familiar equation for network calculations.

Where Q is in cfm, d is in inches, L is in feet, and DP is in inches of water.

Now we can calculate the viscosity and density of air at sea level and thus, the Reynolds number and friction factor. Of course at different altitudes, the absolute air pressure must be adjusted. When calculating the Reynolds number, V must be in Ft./sec. so we use
Vs = 3.0557*cfm/(d^2).

For calculating pressure drops in fittings, we use velocity pressure.
Vp = (V/4005)^2, in. w.g., where V is in ft/min. This V = 183.346*Q/(d^2)

We will now use this technique to design a auto/truck exhaust system. This type of system of course is for exhausting hot fumes from engines. As per CAR-MON (Manufacturer of engine exhaust systems), for cars/light trucks, provide 300 cfm/vehicle and for diesel trucks, provide a minimum of 450 cfm/vehicle. The pickup hose should have a velocity of 3400 ft/min (minimum) and the main duct should have a minimum flow velocity of 2500 ft/min. CAR-MON recommends a 4" diameter pickup for each vehicle.

As for temperature, auto exhaust ranges from 200 – 300 Deg.F and diesel exhaust ranges from 400-500 Deg.F (normal operating rpm). However, for dynamometer tests, diesel exhaust can reach 1000 Deg.F. For this example, we will use a min range temperature of 450 Deg.F. If the fan can handle this flow, higher temperatures should be no problem, as the exhaust density goes down and so does the Hp requirements.

Now, a purist might object saying that the hot exhaust (with hydrocarbons) does not have the same density or viscosity of hot air and they are correct. The author is of the opinion that using air is easily close enough to size and select an exhaust fan. Besides, how do you determine the exact mixture density and viscosity of various engines at differing conditions?

For this example, data was extracted from an actual shop drawing of a CAR-MON exhaust system.

$D_1 = 4"$ $L_1 = 14'$
$D_2 = 4"$ $L_2 = 3'$
$D_3 = 6"$ $L_3 = 20'$
$D_4 = 4"$ $L_4 = 2'$
$D_5 = 8"$ $L_5 = 22'$
$D_6 = 4"$ $L_6 = 2.17'$
$D_7 = 4"$ $L_7 = 18.5'$
$D_8 = 6"$ $L_8 = 17'$
$D_9 = 4"$ $L_9 = 2'$
$D_{10} = 8"$ $L_{10} = 16.75'$
$D_{11} = 10"$ $L_{11} = 41'$

Fittings will be accounted for by using the C0 coefficients multiplied by the velocity pressure entering the fitting.

We will assume also that only cars and light trucks are used, so minimum cfm/vehicle = 300 cfm.

Absolute roughness = 0.00015
P=14.7

Note that we do not put guesses for the flows in the mains, just the branch flows. The nodal equations will take care of the main flow rates. Q1 is given as 300 cfm, minimum flow rate for a branch pickup.

Note also the loss coefficients (Cx) used at each fitting. These come from Industrial Ventilation.

RESULTS:
From results, it can be seen that all pickups have a flow velocity >= 3400 ft/min, and all main flows are >= 2500 ft/min. Also, all pickups are > 300 cfm. Thus, the design criteria is satisfied.

The reader can play with this problem, by changing air temp, duct diameters, duct lengths, fittings, inputing a different value for Q1, inputting the total pressure at fan, total fan cfm, etc. and solving for the rest. TK Solver does not care. You can even input a certain duct velocity at a main and it will solve for the rest.

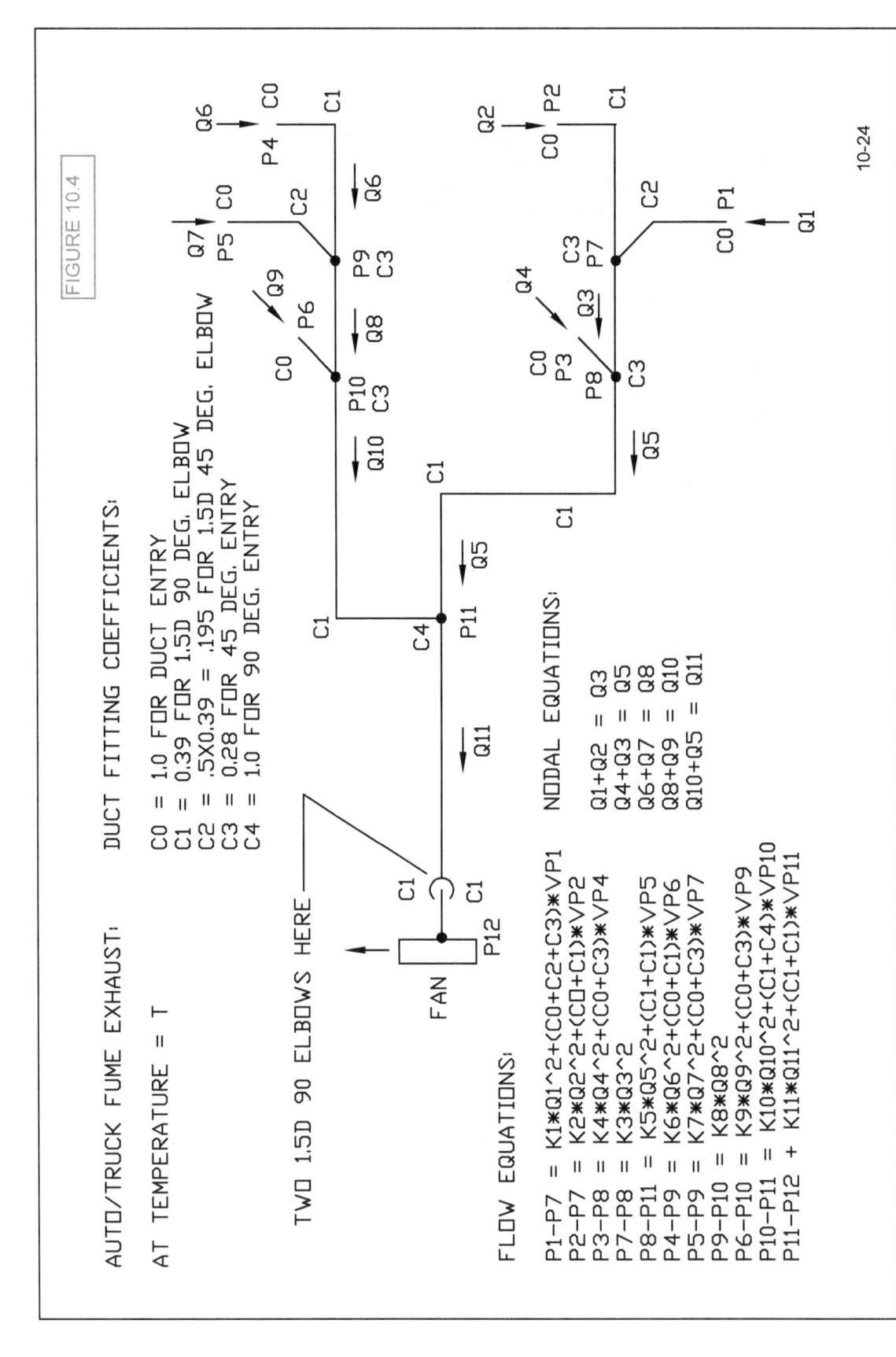

FIGURE 10.4

Rule
;TK SOLVER, DUCT EXHAUSTS
;GARAGE ENGINE EXHAUST

;COMPUTE AIR VISCOSITY (MU)
mu=(8.1852/(0.555*(T+459.2)+120))*((T+459.2)/560)^1.5

;COMPUTE AIR DENSITY (RHO)
rho= (2.6983*P)/(459.2+T)

;COMPUTE AIR VELOCITY, FT/SEC
V1S=3.0557*Q1*(1/D1)^2
V2S=3.0557*Q2*(1/D2)^2
V3S=3.0557*Q3*(1/D3)^2
V4S=3.0557*Q4*(1/D4)^2
V5S=3.0557*Q5*(1/D5)^2
V6S=3.0557*Q6*(1/D6)^2
V7S=3.0557*Q7*(1/D7)^2
V8S=3.0557*Q8*(1/D8)^2
V9S=3.0557*Q9*(1/D9)^2
V10S=3.0557*Q10*(1/D10)^2
V11S=3.0557*Q11*(1/D11)^2

;COMPUTE VM, AIR VELOCITY IN FT/MIN
V1M=V1S*60
V2M=V2S*60
V3M=V3S*60
V4M=V4S*60
V5M=V5S*60
V6M=V6S*60
V7M=V7S*60
V8M=V8S*60
V9M=V9S*60
V10M=V10S*60
V11M=V11S*60

;COMPUTE REYNOLDS NUMBERS
R1=123.9*D1*V1S*rho/mu
R2=123.9*D2*V2S*rho/mu
R3=123.9*D3*V3S*rho/mu
R4=123.9*D4*V4S*rho/mu
R5=123.9*D5*V5S*rho/mu
R6=123.9*D6*V6S*rho/mu
R7=123.9*D7*V7S*rho/mu
R8=123.9*D8*V8S*rho/mu
R9=123.9*D9*V9S*rho/mu
R10=123.9*D10*V10S*rho/mu
R11=123.9*D11*V11S*rho/mu

; COMPUTE VELOCITY PRESSURES IN. W.G.

Rule
VP1=(V1M/4005)^2
VP2=(V2M/4005)^2
VP3=(V3M/4005)^2
VP4=(V4M/4005)^2
VP5=(V5M/4005)^2
VP6=(V6M/4005)^2
VP7=(V7M/4005)^2
VP8=(V8M/4005)^2
VP9=(V9M/4005)^2
VP10=(V10M/4005)^2
VP11=(V11M/4005)^2

;COMPUTE FRICTION FACTOR
f1=(-2*log((e*12/D1)/3.7-(5.02/R1)*log((e*12/D1)/3.7+14.5/R1)))^-2
f2=(-2*log((e*12/D2)/3.7-(5.02/R2)*log((e*12/D2)/3.7+14.5/R2)))^-2
f3=(-2*log((e*12/D3)/3.7-(5.02/R3)*log((e*12/D3)/3.7+14.5/R3)))^-2
f4=(-2*log((e*12/D4)/3.7-(5.02/R4)*log((e*12/D4)/3.7+14.5/R4)))^-2
f5=(-2*log((e*12/D5)/3.7-(5.02/R5)*log((e*12/D5)/3.7+14.5/R5)))^-2
f6=(-2*log((e*12/D6)/3.7-(5.02/R6)*log((e*12/D6)/3.7+14.5/R6)))^-2
f7=(-2*log((e*12/D7)/3.7-(5.02/R7)*log((e*12/D7)/3.7+14.5/R7)))^-2
f8=(-2*log((e*12/D8)/3.7-(5.02/R8)*log((e*12/D8)/3.7+14.5/R8)))^-2
f9=(-2*log((e*12/D9)/3.7-(5.02/R9)*log((e*12/D9)/3.7+14.5/R9)))^-2
f10=(-2*log((e*12/D10)/3.7-(5.02/R10)*log((e*12/D10)/3.7+14.5/R10)))^-2
f11=(-2*log((e*12/D11)/3.7-(5.02/R11)*log((e*12/D11)/3.7+14.5/R11)))^-2

;COMPUTE DUCT CONSTANTS
K1=0.3346*f1*L1*rho*(1/D1)^5
K2=0.3346*f2*L2*rho*(1/D2)^5
K3=0.3346*f3*L3*rho*(1/D3)^5
K4=0.3346*f4*L4*rho*(1/D4)^5
K5=0.3346*f5*L5*rho*(1/D5)^5
K6=0.3346*f6*L6*rho*(1/D6)^5
K7=0.3346*f7*L7*rho*(1/D7)^5
K8=0.3346*f8*L8*rho*(1/D8)^5
K9=0.3346*f9*L9*rho*(1/D9)^5
K10=0.3346*f10*L10*rho*(1/D10)^5
K11=0.3346*f11*L11*rho*(1/D11)^5

;LINK EQUATIONS
P1-P7=K1*Q1^2+(C0+C1+C3)*VP1
P2-P7=K2*Q2^2+C1*VP2
P3-P8=K4*Q4^2+(C0+C3)*VP4
P7-P8=K3*Q3^2
P8-P11=K5*Q5^2+(C1+C1)*VP5
P4-P9=K6*Q6^2+(C0+C1)*VP6
P5-P9=K7*Q7^2+(C0+C3)*VP7
P9-P10=K8*Q8^2
P6-P10=K9*Q9^2+(C0+C3)*VP9

Rule
P11-P12=K11*Q11^2+(C1+C1)*VP11
P10-P11=K10*Q10^2+(C1+C4)*VP10

;NODAL EQUATIONS
Q1+Q2=Q3
Q3+Q4=Q5
Q7+Q6=Q8
Q9+Q8=Q10
Q10+Q5=Q11

Variables

Status	Input	Name	Output	Unit	Comment
		R1			Reynolds No. Duct 1, typical
	300	Q1		CFM	FLOW RATE FOR PICKUP AT P1, GIVEN
	4	D1		INCHES	DIAMETER FOR DUCT 1, GIVEN
	.00015	e			RELATIVE ROUGHNESS
		f1			FRICTION FACTOR, DUCT 1, TYPICAL
	20	L1		FEET	LENGTH OF DUCT 1, TYPICAL
		mu		CENTIPOISE	AIR VISCOSITY
		R2			
	4	D2		INCHES	DIAMETER FOR DUCT 2, GIVEN
		R3			
	6	D3		INCHES	
		R4			
	4	D4		INCHES	
		R5			
	8	D5		INCHES	
		R6			
	4	D6		INCHES	
		R7			
	4	D7		INCHES	
		R8			
	6	D8		INCHES	
Guess	1	Q2		CFM	CALCULATED FLOW AT P2
		Q3		CFM	CALCULATED FLOW IN MAIN
Guess	1	Q4		CFM	CALCULATED FLOW AT P3
		Q5		CFM	CALCULATED FLOW IN MAIN
Guess	1	Q6		CFM	CALCULATED FLOW AT P4
Guess	1	Q7		CFM	CALCULATED FLOW AT P5
		Q8		CFM	CALCULATED FLOW IN MAIN
		f2			
		f3			
		f4			
		f5			
		f6			
		f7			
		f8			
	3	L2		FEET	
	20	L3		FEET	

Status	Input	Name	Output	Unit	Comment
	2	L4		FEET	
	22	L5		FEET	
	2.17	L6		FEET	
	18.5	L7		FEET	
	17	L8		FEET	
	0	P1		"W.G.	INLET ATMO PRESSURE AT P1, TYPICAL
	0	P2		"W.G.	
	0	P4		"W.G.	
	0	P3		"W.G.	
	0	P6		"W.G.	
	0	P5		"W.G.	
Guess	-1	P8		"W.G.	CALCULATED PRESSURE
Guess	-1	P7		"W.G.	CALCULATED PRESSURE
		rho		LB/CU. FT.	AIR DENSITY AT TEMP T
	450	T		DEG.F	AIR TEMPERATURE
		VP1		"W.G.	VELOCITY PRESSURE, DUCT 1, TYPICAL
		VP2			
		VP3			
		VP4			
		VP5			
		VP6			
		VP7			
		VP8			
Guess	-1	P9		IN. W.G.	CALCULATED PRESSURE
		V1M		FT/MIN	DUCT AIR VELOCITY IN PICKUP
		V2M		FT/MIN	DUCT AIR VELOCITY IN PICKUP
		V3M		FT/MIN	DUCT AIR VELOCITY IN MAIN
		V4M		FT/MIN	DUCT AIR VELOCITY IN PICKUP
		V5M		FT/MIN	DUCT AIR VELOCITY IN MAIN
		V6M		FT/MIN	DUCT AIR VELOCITY IN PICKUP
		V7M		FT/MIN	DUCT AIR VELOCITY IN PICKUP
		V8M		FT/MIN	DUCT AIR VELOCITY IN MAIN
		f9			
	2	L9			
	4	D9			
		f10			
	16.75	L10			

Variables

Status	Input	Name	Output	Unit	Comment
	8	D10			
		f11			
	41	L11			
	10	D11			
	.195	C2			LOSS COEFFICIENT FOR 1.5D 45 DEG ELBOW
	.28	C3			LOSS COEFFICIENT FOR 45 DEG. ENTRY
	1	C0			LOSS COEFFICIENT FOR AIR ENTERING DUCT
	.39	C1			LOSS COEFFICIENT FOR 1.5D 90 DEG. ELBOW
Guess	-1	P11		"W.G.	CALCULATED PRESSURE
Guess	-1	P10		"W.G.	CALCULATED PRESSURE
Guess	-1	P12		"W.G.	CALCULATED PRESSURE AT FAN
Guess	1	Q9		CFM	CALCULATED FLOW AT P6
		VP9			
	1	Q10		CFM	CALCULATED FLOW IN MAIN
		C4			
		VP10			
		Q11		CFM	TOTAL AIR FLOW AT FAN
	14.7	P		PSI	ATMOSPHERIC PRESSURE
		V9M		FT/MIN	FLOW VELOCITY IN PICKUP
		V10M		FT/MIN	FLOW VELOCITY IN MAIN
		V11M		FT/MIN	FLOW VELOCITY IN MAIN
		R9			
		R10			
		R11			
		VP11			
		V1S		FT/SEC	AIR FLOW VELOCITY, TYPICAL
		V2S			
		V3S			
		V4S			
		V5S			
		V6S			
		V7S			
		V8S			
		V9S			
		V10S			
		V11S			
		K1			DUCT COEFFICIENT, TYPICAL

Status	Input	Name	Output	Unit	Comment
		K2			
		K3			
		K4			
		K5			
		K6			
		K7			
		K8			
		K9			
		K10			
		K11			

Variables GARAGE ENGINE EXHAUST.tkw

Status	Input	Name	Output	Unit	Comment
		R1			Reynolds No. Duct 1, typical
	300	Q1	45694.2108271494	CFM	FLOW RATE FOR PICKUP AT P1, GIVEN
	4	D1		INCHES	DIAMETER FOR DUCT 1, GIVEN
	.00015	e			RELATIVE ROUGHNESS
		f1	.0228270119408257		FRICTION FACTOR, DUCT 1, TYPICAL
	20	L1		FEET	LENGTH OF DUCT 1, TYPICAL
		mu	.0271100453377478	CENTIPOISE	AIR VISCOSITY
		R2	101964.372835647		
	4	D2		INCHES	DIAMETER FOR DUCT 2, GIVEN
		R3	98439.0557751973	INCHES	
	6	D3		INCHES	
		R4	72445.5325302673	INCHES	
	4	D4		INCHES	
		R5	110052.058096532	INCHES	
	8	D5		INCHES	
		R6	62738.6041390963	INCHES	
	4	D6		INCHES	
		R7	53892.8956078646	INCHES	
	4	D7		INCHES	
		R8	77754.3331646406	INCHES	
	6	D8		INCHES	
		Q2	669.43517126067	CFM	CALCULATED FLOW AT P2
		Q3	969.43517126067	CFM	CALCULATED FLOW IN MAIN
		Q4	475.632675686064	CFM	CALCULATED FLOW AT P3
		Q5	1445.06784694673	CFM	CALCULATED FLOW IN MAIN
		Q6	411.902971974427	CFM	CALCULATED FLOW AT P4
		Q7	353.827506585433	CFM	CALCULATED FLOW AT P5
		Q8	765.730478859859	CFM	CALCULATED FLOW IN MAIN
		f2	.0201076140816942		
		f3	.0195733959009406		
		f4	.0211343253670966		
		f5	.0188920795023670		
		f6	.0216230817276315		
		f7	.0221779234185783		
		f8	.0203358751795905		
	3	L2		FEET	
	20	L3		FEET	

Variables   GARAGE ENGINE EXHAUST.tkw

Status	Input	Name	Output	Unit	Comment
	2	L4		FEET	
	22	L5		FEET	
	2.17	L6		FEET	
	18.5	L7		FEET	
	17	L8		FEET	
	0	P1		"W.G.	INLET ATMO PRESSURE AT P1, TYPICAL
	0	P2		"W.G.	
	0	P4		"W.G.	
	0	P3		"W.G.	
	0	P6		"W.G.	
	0	P5		"W.G.	
		P8	-2.5067432499427	"W.G.	CALCULATED PRESSURE
		P7	-1.8161037531632B	"W.G.	CALCULATED PRESSURE
		rho	.0436262758468984	LB/CU. FT.	AIR DENSITY AT TEMP T
	450	T		DEG.F	AIR TEMPERATURE
		VP1	.7367521848567	"W.G.	VELOCITY PRESSURE, DUCT 1, TYPICAL
		VP2	3.6685629425213S		
		VP3	1.5196774902425S		
		VP4	1.8519202838742T		
		VP5	1.0684037811340S		
		VP6	1.3888929517769T		
		VP7	1.0248542512174T		
		VP8	.948125340961631		
		P9	-2.0440445563418B	IN. W.G.	CALCULATED PRESSURE
		V1M	3437.6625	FT/MIN	DUCT AIR VELOCITY IN PICKUP
		V2M	7670.97394807961	FT/MIN	DUCT AIR VELOCITY IN PICKUP
		V3M	4937.17175470205	FT/MIN	DUCT AIR VELOCITY IN MAIN
		V4M	5450.21537660214	FT/MIN	DUCT AIR VELOCITY IN PICKUP
		V5M	4139.71295617044	FT/MIN	DUCT AIR VELOCITY IN MAIN
		V6M	4719.94466798346	FT/MIN	DUCT AIR VELOCITY IN PICKUP
		V7M	4054.46516952415	FT/MIN	DUCT AIR VELOCITY IN PICKUP
		V8M	3899.73770555894	FT/MIN	DUCT AIR VELOCITY IN MAIN
		f9	.0211189419587618S		
	2	L9			
	4	D9			
		f10	.019393123417611Z		
	16.75	L10			

Variables   GARAGE ENGINE EXHAUST.tkw

Variables

Status	Input	Name	Output	Unit	Comment
	8	D10			
		f11	.0175324878209279		
	41	L11			
	10	D11			
	.195	C2			LOSS COEFFICIENT FOR 1.5D 45 DEG ELBOW
	.28	C3			LOSS COEFFICIENT FOR 45 DEG. ENTRY
	1	C0			LOSS COEFFICIENT FOR AIR ENTERING DUCT
	.39	C1			LOSS COEFFICIENT FOR 1.5D 90 DEG. ELBOW
		P11	-3.7267333328432	"W.G.	CALCULATED PRESSURE
		P10	-2.42456785480442	"W.G.	CALCULATED PRESSURE
		P12	-5.65231870028485	"W.G.	CALCULATED PRESSURE AT FAN
		Q9	467.737010951896	CFM	CALCULATED FLOW AT P6
		VP9	1.79094560258678		
		Q10	1233.46748951176	CFM	CALCULATED FLOW IN MAIN
	1	C4			
		VP10	.778420661640902		
	14.7	Q11	2678.53533645849	CFM	TOTAL AIR FLOW AT FAN
		P		PSI	ATMOSPHERIC PRESSURE
		V9M	5359.73994137141	FT/MIN	FLOW VELOCITY IN PICKUP
		V10M	3533.53744471975	FT/MIN	FLOW VELOCITY IN MAIN
		V11M	4910.88025656972	FT/MIN	FLOW VELOCITY IN MAIN
		R9	71242.9119669888		
		R10	93937.2058569748		
		R11	163191.411162805		
		VP11	1.50353536820337		
		V1S	57.294375	FT/SEC	AIR FLOW VELOCITY, TYPICAL
		V2S	127.849565801327		
		V3S	82.2861959117008		
		V4S	90.8369229433691		
		V5S	68.9952159336174		
		V6S	78.6657444466391		
		V7S	67.5744194920692		
		V8S	64.9956284259823		
		V9S	89.3289990228568		
		V10S	58.8922907453292		
		V11S	81.8480042761621		
		K1	.0000065080844931751		DUCT COEFFICIENT, TYPICAL

GARAGE ENGINE EXHAUST.tkw

Status	Input	Name	Output	Unit	Comment
		K2	.00000008599157711080		
		K3	.00000007348758945786		
		K4	.00000006025491885121		
		K5	.00000001851512242717		
		K6	.00000006688849812880		
		K7	.00000005848799435105		
		K8	.00000006489774620220		
		K9	.00000006041199496928		
		K10	.00000001447060588430		
		K11	.00000001049304366941		

CHAPTER 11 – FANS IN DUCT NETWORK

EXHAUSTS TO A FAN

PARALLEL FANS

Fans can be accounted for in ductwork in the same way that pumps were modeled in piping systems. A fan is selected, then a curve fit for the fan's performance is done. This is inserted into the equations to represent the pressure of the fan depending on the CFM.

The problem will be the 7 exhausts problem from Chapter 10, but with a fan added at the end of the duct. The fan selected is a Greenheck roof mounted unit, GB-220HP, belt drive. Results show that with an assumed Q1=420, the final discharge pressure of the fan is approximately 0.14"w.g. Total CFM = approx. 5089 CFM. Note that this matches the fan performance to the system curve. See Figures 11.1.

Fans in series is usually not seen in practice, though there is no real reason not to use them that way, The fan selected may have just the right volume, but not enough pressure. A second identical fan in line would provide the necessary pressure boost. The first fan discharge would simply be the suction of the second fan, to act as a booster fan. This can be left to the reader.

PROBLEM 11.1 – 7 EXHAUSTS TO A FAN

PROBLEM 11.2 – PARALLEL FANS

Fans in parallel are fairly common, especially for multiple toilet ceiling fans in apartments and hotels. For stacked toilets, a common vertical riser is between the bathrooms. The discharge from the ceiling fans discharge into the common riser. Of course, each fan has a backdraft damper to prevent discharge from another toilet fan from coming into its housing. This problem will consider (3) parallel Greenheck CSP-B110 direct drive ceiling exhaust fan. Note that duct entry loss at a 90 Deg. Angle (perpendicular) to the exhaust main = 1*Vp, or just Vp. Also note that these fans are considered to all be at the same level. This problem is based upon an actual project with several toilets in a office building. This could of course be expanded with several fans on each floor discharging into a common riser for 2 or more floors up to the roof. It would work, providing the fans selected have sufficient static pressure. What this problem illustrates is merely the technique.

You may notice that these fans in parallel resemble the condensate pumps in parallel. Different fluid, same technique.

Note that when applying fans in parallel, to stay out of the unstable region of the fan curve.

GREENHECK GB-220HP ROOF MOUNTED EXHAUST FAN
1230 RPM, 3HP

4th Degree Polynomial Fit: y=a+bx+cx^2+dx^3...
Coefficient Data:
a = -1.60199915370E-006
b = 2.45725626706E-003
c = -7.23018740136E-007
d = 8.50049927127E-011
e = -4.29943507220E-015

Standard Error: 0.0041912
Correlation Coefficient: 0.9999916

7 EXHAUSTS WITH FAN

FIG. 11.1

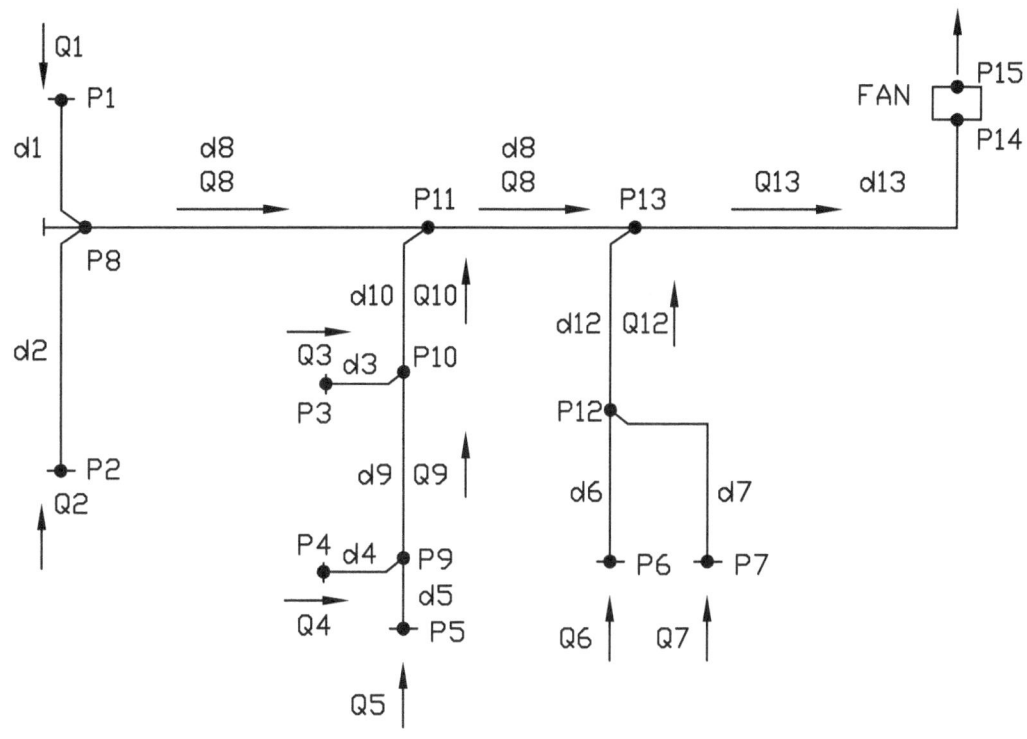

DUCT LINKS:

P1-P8=K1*Q1^K+1.18*VP1
P2-P8=K2*Q2^K+1.18*VP2
P3-P10=K3*Q3^K+1.18*VP3
P4-P9=K4*Q4^K+1.18*VP4
P8-P11=K8*Q8^K
P5-P9=K5*Q5^K+VP5
P4-P9=K4*Q4^K+1.18*VP4
P10-P11=K10*Q19^K+0.18*VP10
P11-P13=K11*Q11^K
P6-P12=K6*Q6^K+VP6
P7-P12=K7*Q7^K+1.18*VP7
P12-P13=K12*Q12^K+0.18*VP12
P13-P14=K13*Q13^K
P15=P14+A+B*Q13+C*Q13^2+D*Q13^3+E*Q13^4

NODAL EQUATIONS:

Q1+Q2=Q8
Q4+Q5=Q9
Q3+Q9=Q10
Q8+Q10=Q11
Q6+Q7=Q12
Q12+Q11=Q13

GIVEN PRESSURES:

P1=0   P5=0
P2=0   P6=0
P3=0   P7=0
P4=0

SET Q1 = 420

11-4

Rule

;>>>/* DUCT LENGTH CONSTANTS 8/20/14 */

K1=1.23174*10^-3*L1*(1/d1)^4.51
K2=1.23174*10^-3*L2*(1/d2)^4.51
K3=1.23174*10^-3*L3*(1/d3)^4.51
K4=1.23174*10^-3*L4*(1/d4)^4.51
K5=1.23174*10^-3*L5*(1/d5)^4.51
K6=1.23174*10^-3*L6*(1/d6)^4.51
K7=1.23174*10^-3*L7*(1/d7)^4.51
K8=1.23174*10^-3*L8*(1/d8)^4.51
K9=1.23174*10^-3*L9*(1/d9)^4.51
K10=1.23174*10^-3*L10*(1/d10)^4.51
K11=1.23174*10^-3*L11*(1/d11)^4.51
K12=1.23174*10^-3*L12*(1/d12)^4.51
K13=1.23174*10^-3*L13*(1/d13)^4.51

;>>>/* AREAS */
A1=0.785*(d1/12)^2
A2=0.785*(d2/12)^2
A3=0.785*(d3/12)^2
A4=0.785*(d4/12)^2
A5=0.785*(d5/12)^2
A6=0.785*(d6/12)^2
A7=0.785*(d7/12)^2
A8=0.785*(d8/12)^2
A9=0.785*(d9/12)^2
A10=0.785*(d10/12)^2
A11=0.785*(d11/12)^2
A12=0.785*(d12/12)^2
A13=0.785*(d13/12)^2

;>>>/* VELOCITIES */
V1=Q1/A1
V2=Q2/A2
V3=Q3/A3
V4=Q4/A4
V5=Q5/A5
V6=Q6/A6
V7=Q7/A7
V8=Q8/A8
V9=Q9/A9
V10=Q10/A10
V11=Q11/A11
V12=Q12/A12
V13=Q13/A13

;>>>/* VELOCITY PRESSURES */
VP1=(V1/4005)^2
VP2=(V2/4005)^2

Rule
VP3=(V3/4005)^2
VP4=(V4/4005)^2
VP5=(V5/4005)^2
VP6=(V6/4005)^2
VP7=(V7/4005)^2
VP8=(V8/4005)^2
VP9=(V9/4005)^2
VP10=(V10/4005)^2
VP11=(V11/4005)^2
VP12=(V12/4005)^2
VP13=(V13/4005)^2

;>>>/* LINKS */
P1-P8=K1*Q1^K+1.18*VP1
P2-P8=K2*Q2^K+1.18*VP2
P8-P11=K8*Q8^K
P5-P9=K5*Q5^K+VP5
P4-P9=K4*Q4^K+1.18*VP4
P9-P10=K9*Q9^K
P3-P10=K3*Q3^K+1.18*VP3
P10-P11=K10*Q10^K+0.18*VP10
P11-P13=K11*Q11^K
P6-P12=K6*Q6^K+VP6
P7-P12=K7*Q7^K+1.18*VP7
P12-P13=K12*Q12^K+.18*VP12
P13-P14=K13*Q13^K
P15=P14+A+B*Q13+C*Q13^2+D*Q13^3+E*Q13^4

;>>>/* NODAL EQUATIONS */
Q1+Q2=Q8
Q4+Q5=Q9
Q3+Q9=Q10
Q8+Q10=Q11
Q6+Q7=Q12
Q11+Q12=Q13

Status	Input	Name	Output	Unit	Comment
	15	K1			
		L1		FEET	LENGTH OF DUCT 1, TYPICAL
	6	d1		inches	DIAMETER OF DUCT 1, TYPICAL
		K2			
	18	L2			
	6	d2		inches	
		K3			
	14	L3			
	7	d3		inches	
		K4			
	11	L4			
	8	d4		inches	
		K5			
	4	L5			
	7	d5		inches	
		K6			
	19	L6			
	11	d6		inches	
		K8			
	21	L8			
	14	d8		inches	
		A1		SQ.FT.	FLOW AREA OF DUCT 1, TYPICAL
		A2			
		A3			
		A4			
		A5			
		A6			
		A7			
	12	d7		inches	
		A8			
		V1		FPM	AIR VELOCITY
	420	Q1		cfm	
		V2		FPM	AIR VELOCITY
Guess	1	Q2		cfm	
		V3		FPM	
Guess	1	Q3		cfm	
		V4		FPM	

Variables   7 exhausts with faN.tkw

Status	Input	Name	Output	Unit	Comment
Guess	1	Q4		cfm	
		V5		FPM	
Guess	1	Q5		cfm	
		V6		FPM	
Guess	1	Q6		cfm	
		V7		FPM	
Guess	1	Q7		cfm	
		V8		FPM	
Guess	1	Q8		cfm	
		VP1		in. w.g.	VELOCITY PRESSURE
		VP2			
		VP3			
		VP4			
		VP5			
		VP6			
		VP7			
		VP8			
	0	P1			
	0	P5			
	1.82	K			
	0	P6			
	0	P2			
	0	P7			
	0	P3			
Guess	1	P8		IN. W.G.	
	0	K7			
Guess	1	P4		in. w.g.	
	20	P9			
		L7			
	28	K9			
	12	L9			
		d9			
	22	K10			
	14	L10			
		d10			
		K11			
	21	L11			

Variables

7 exhausts with faN.tkw

Status	Input	Name	Output	Unit	Comment
	13	d11			
		K12			
	22	L12			
	12	d12			
		K13			
	34	L13			
	16	d13			
		A9			
		A10			
		A11			
		A12			
		A13			
		V9			
Guess	1	Q9			
Guess	1	V10			
Guess	1	Q10			
Guess	1	V11			
Guess	1	Q11			
Guess	1	V12			
Guess	1	Q12			
Guess	1	V13			
Guess	1	Q13		CFM	TOTAL AIR FLOW
		VP9			
		VP10			
		VP11			
		VP12			
		VP13			
Guess	1	P11			
Guess	1	P10			
Guess	1	P13			
Guess	1	P12			
Guess	1	P14		IN. W.G.	SUCTION PRESSURE AT FAN, CALCULATED
	-.0000016019991Ƽ	A			CURVE COEFFICIENT FOR GREENHECK GB-220HP
	.0024572556865Ƽ	B			CURVE COEFFICIENT FOR GREENHECK GB-220HP
	-7.2301874013ƼE-7	C			CURVE COEFFICIENT FOR GREENHECK GB-220HP
	8.500499271Ƽ7E-1Ƽ	D			CURVE COEFFICIENT FOR GREENHECK GB-220HP

Variables
7 exhausts with faN.tkw

Status	Input	Name	Output	Unit	Comment
	-4.2943434266E-1	E			CURVE COEFFICIENT FOR GREENHECK GB-220HP
		P15	.1385973808420	IN. W.G.	DISCHARGE PRESSURE AT FAN, CALCULATED

Variables

7 exhausts with faN.tkw

Status	Input	Name	Output	Unit	Comment
	15	K1	5.71673636927E-(	FEET	LENGTH OF DUCT 1, TYPICAL
	6	L1			
		d1	6.86008364312E-(	inches	DIAMETER OF DUCT 1, TYPICAL
	18	K2			
	6	L2			
		d2	2.66228729924E-(	inches	
	14	K3			
	7	L3			
		d3	1.145449334685E-(	inches	
	11	K4			
	8	L4			
		d4	7.606351407E-7	inches	
	4	K5			
	7	L5			
		d5	4.7053467791E-7	inches	
	19	K6			
	11	L6			
		d6	1.75267301023E-;	inches	
	21	K8			
	14	L8			
		d8	?.19625	inches	
		A1	.19625	SQ.FT.	FLOW AREA OF DUCT 1, TYPICAL
		A2	.2671180555555(		
		A3	.348888888888888		
		A4	.2671180555555(		
		A5	.6596180555555(		
		A6	.785		
		A7	1.06847222222222	inches	
	12	A8			
		V1	?2140.127388535	FPM	AIR VELOCITY
	420	Q1		cfm	
		V2	2035.04773296773	FPM	AIR VELOCITY
		Q2	399.37811759483.	cfm	
		V3	2004.7813580398	FPM	
		Q3	535.51329817362	cfm	
		V4	1750.1997428553	FPM	

Variables   7 exhausts with faN.tkw

Status	Input	Name	Output	Unit	Comment
		Q4	610.6252436184	cfm	
		V5	2125.5073040612	FPM	
		Q5	567.7613781299		
		V6	1808.6819282601	FPM	
		Q6	1193.0392566374		
		V7	1735.7420134253	FPM	
		Q7	1362.5574805389		
		V8	766.8688998677	FPM	
		Q8	819.3781175948	cfm	
		VP1	.28554476936025	in. w.g.	VELOCITY PRESSURE
		VP2	.2581928192415		
		VP3	.2505699519510		
		VP4	.1909722173058		
		VP5	.2816567492642		
		VP6	.2039479562915		
		VP7	.1878301522079		
		VP8	.0366637776178		
	0	P1			
	0	P5			
	1.82	K			
	0	P6			
	0	P2			
	0	P7			
	0	P3			
		P8	-.6769294850370	IN. W.G.	
		K7	3.3453504582E-		
	0	P4			
		P9	-.3599563452708	in. w.g.	
	20	L7			
		K9	4.6834906415E-		
	28	L9			
	12	d9			
		K10	1.8361336297E-		
	22	L10			
	14	d10			
		K11	2.4482390509E-		
	21	L11			

Variables 7 exhausts with faN.tkw

Status	Input	Name	Output	Unit	Comment
	13	d11	3.6798850406E-		
		K12			
	22	L12			
	12	d12			
		K13	1.5538777245E-		
	34	L13			
	16	d13			
		A9	.785		
		A10	1.0684722222222		
		A11	.9212847222222		
		A12	.785		
		A13	1.3955555555555		
		V9	1501.1294544565		
		Q9	1178.3866217483		
		V10	1604.0659591106		
		Q10	1713.899919922		
		V11	2749.7232684010		
		Q11	2533.2780375168		
		V12	3255.5372448106		
		Q12	2555.5967371763		
		V13	3646.4867016113		
		Q13	5088.8747746932	CFM	TOTAL AIR FLOW
		VP9	.14048541938288		
		VP10	.16041294207319		
		VP11	.47138193692254		
		VP12	.66075475271075		
		VP13	.82898033295013		
		P11	-.71210515757579		
		P10	-.54205674346850		
		P13	-1.09541955805252		
		P12	-.39106038167430		
		P14	-1.9613192707231	IN. W.G.	SUCTION PRESSURE AT FAN, CALCULATED

Variables

-.00000160199915	A		CURVE COEFFICIENT FOR GREENHECK GB-220HP
.00245725568652	B		CURVE COEFFICIENT FOR GREENHECK GB-220HP
-7.2301874013GE-7	C		CURVE COEFFICIENT FOR GREENHECK GB-220HP
8.50049927127E-1	D		CURVE COEFFICIENT FOR GREENHECK GB-220HP

7 exhausts with faN.tkw

Status	Input	Name	Output	Unit	Comment
	-4.2994343426 6E-1	E			CURVE COEFFICIENT FOR GREENHECK GB-220HP
		P15	.13859336294456	IN. W.G.	DISCHARGE PRESSURE AT FAN, CALCULATED

7 exhausts with faN.tkw

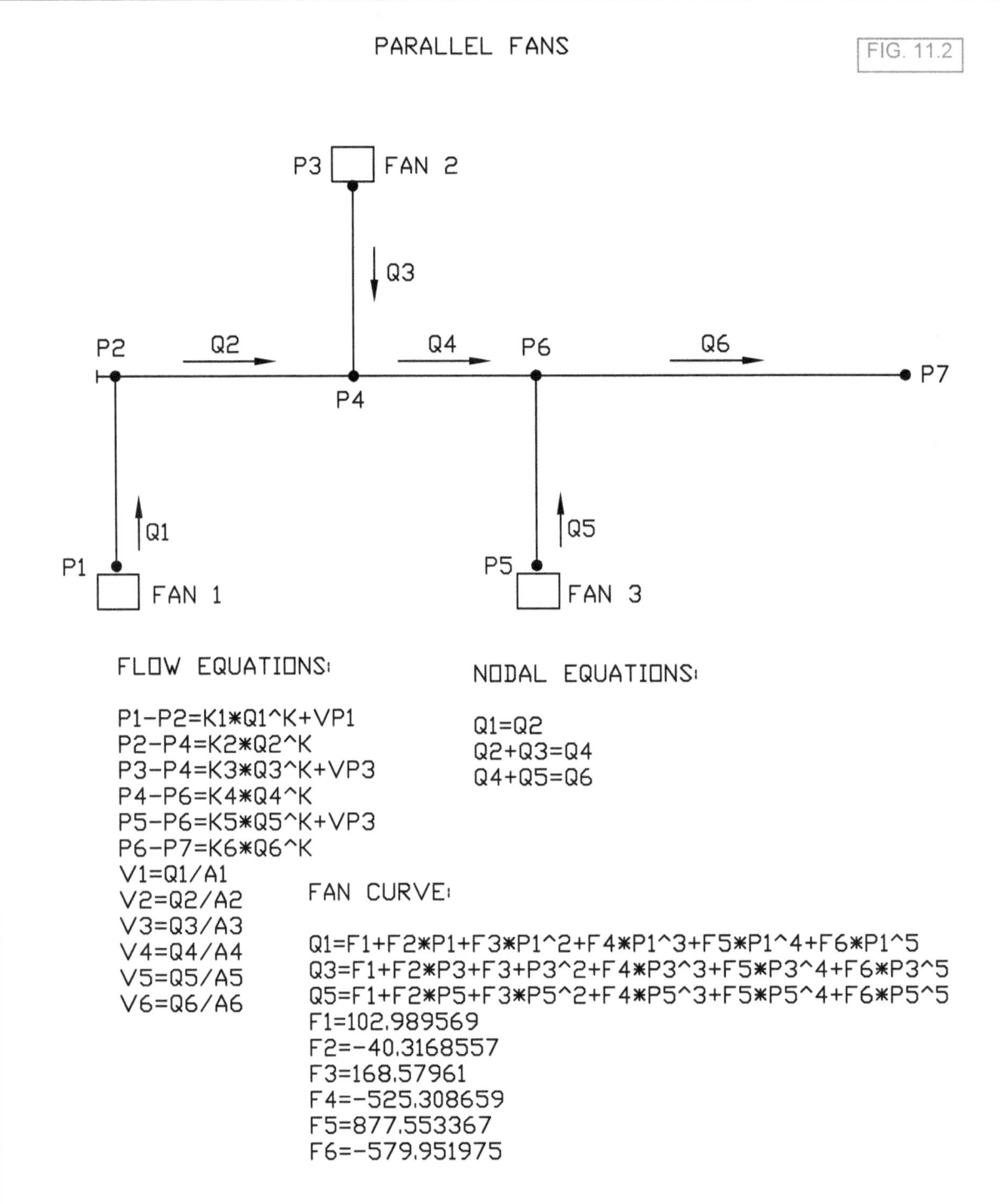

PARALLEL FANS   FIG. 11.2

FLOW EQUATIONS:

P1-P2=K1*Q1^K+VP1
P2-P4=K2*Q2^K
P3-P4=K3*Q3^K+VP3
P4-P6=K4*Q4^K
P5-P6=K5*Q5^K+VP3
P6-P7=K6*Q6^K
V1=Q1/A1
V2=Q2/A2
V3=Q3/A3
V4=Q4/A4
V5=Q5/A5
V6=Q6/A6

NODAL EQUATIONS:

Q1=Q2
Q2+Q3=Q4
Q4+Q5=Q6

FAN CURVE:

Q1=F1+F2*P1+F3*P1^2+F4*P1^3+F5*P1^4+F6*P1^5
Q3=F1+F2*P3+F3+P3^2+F4*P3^3+F5*P3^4+F6*P3^5
Q5=F1+F2*P5+F3*P5^2+F4*P5^3+F5*P5^4+F6*P5^5
F1=102.989569
F2=-40.3168557
F3=168.57961
F4=-525.308659
F5=877.553367
F6=-579.951975

11-15

GREENHECK CSP-B110 CEILING FAN

Polynomial Regression Worksheet ($y = b_1 + b_2 \cdot x + b_3 \cdot x^2 + \ldots$)

	X	Y	residuals		SUMMARY STATS
1	0	103	+1.043E-2	order	5
2	.1	100	−2.003E-1	N	8
3	.125	100	+2.454E-1	Syx	.241901139
4	.25	98	−1.002E-1	adj R2	.997791843
5	.375	97	+7.140E-2	p	.010016335
6	.5	96	−3.605E-2	b1	102.989569
7	.625	94	+1.075E-2	b2	−40.3168557
8	.75	86	−1.416E-3	b3	168.579610
9				b4	−525.308659
10				b5	877.553367
11				b6	−579.951975
12				b7	0
13				b8	0
14				b9	0

Rule
;>>>/* TK SOLVER */
;>>>/* 3 PARALLEL EXHAUST FANS */
;>>>/* 08/21/14 */

;>>>/* FLOW EQUATIONS */
P1-P2=K1*Q1^K+VP1
P2-P4=K2*Q2^K
P3-P4=K3*Q3^K+VP3
P4-P6=K4*Q4^K
P5-P6=K5*Q5^K+VP5
P6-P7=K6*Q6^K

;>>>/* DUCT K FACTORS */
K1=1.23174*10^-3*L1*(1/D1)^4.51
K2=1.23174*10^-3*L2*(1/D2)^4.51
K3=1.23174*10^-3*L3*(1/D3)^4.51
K4=1.23174*10^-3*L4*(1/D4)^4.51
K5=1.23174*10^-3*L5*(1/D5)^4.51
K6=1.23174*10^-3*L6*(1/D6)^4.51

;>>>/* VELOCITIES */
V1=Q1/A1
V2=Q2/A2
V3=Q3/A3
V4=Q4/A4
V5=Q5/A5
V6=Q6/A6

A1=0.785*(D1/12)^2
A2=0.785*(D2/12)^2
A3=0.785*(D3/12)^2
A4=0.785*(D4/12)^2
A5=0.785*(D5/12)^2
A6=0.785*(D6/12)^2

VP1=(V1/4005)^2
VP2=(V2/4005)^2
VP3=(V3/4005)^2
VP4=(V4/4005)^2
VP5=(V5/4005)^2
VP6=(V6/4005)^2
;>>>/* NODAL EQUATIONS */
Q1=Q2
Q2+Q3=Q4
Q4+Q5=Q6
;>>>/* FAN CURVE */
Q1=F1+F2*P1+F3*P1^2+F4*P1^3+F5*P1^4+F6*P1^5
Q3=F1+F2*P3+F3*P3^2+F4*P3^3+F5*P3^4+F6*P3^5
Q5=F1+F2*P5+F3*P5^2+F4*P5^3+F5*P5^4+F6*P5^5

Status	Input	Name	Output	Unit	Comment
Guess	1	P1			
Guess	1	P2			
		K1			
Guess	1	Q1		CFM	AIR FLOW FROM FAN 1
	1.82	K			
		V1		FT/MIN	AIR VELOCITY, DUCT 1, TYPICAL
Guess	1	P4		IN. W.G.	PRESSURE IN MAIN, FROM FAN 1
		K2			
Guess	1	Q2		CFM	AIR FLOW FROM FAN 1
		V2			
		P3			
		K3			
Guess	1	Q3		CFM	AIR FLOW FROM FAN 2
		V3			
Guess	1	P6			
		K4			
Guess	1	Q4		CFM	AIR FLOW FROM FAN 1 + FAN 2
		V4			
Guess	1	P5			
		K5			
Guess	1	Q5		CFM	AIR FLOW FROM FAN 3
		V5			
	.1	P7		IN. W.G.	PRESSURE AT END OF DUCT RUN
		K6			
Guess	1	Q6		CFM	TOTAL AIR FLOW
		V6			
	5	L1			
	4	D1			
	8	L2			
	4	D2			
	11	L3			
	4	D3			
	13	L4			
	6	D4			
	12	L5			
	4	D5			
	34	L6			

Variables

11.03 parallel fans CSP B110.tkw

Status	Input	Name	Output	Unit	Comment
	6	D6			
		A1			
		A2			
		A3			
		VP1			
		VP3			
		VP5			
		A4			
		A5			
		A6			
		VP2			
		VP4			
		VP6			
	102.989569	F1			FAN CURVE COEFFICIENT FOR GREENHECK CSP-B110
	-40.3168557	F2			FAN CURVE COEFFICIENT FOR GREENHECK CSP-B110
	168.57961	F3			FAN CURVE COEFFICIENT FOR GREENHECK CSP-B110
	-525.308659	F4			FAN CURVE COEFFICIENT FOR GREENHECK CSP-B110
	877.553367	F5			FAN CURVE COEFFICIENT FOR GREENHECK CSP-B110
	-579.951975	F6			FAN CURVE COEFFICIENT FOR GREENHECK CSP-B110

Variables

Status	Input	Name	Output	Unit	Comment
		P1	.700265459		
		P2	.590199491		
		K1	1.186311E-5		
		Q1	90.4006724	CFM	AIR FLOW FROM FAN 1
	1.82	K			
		V1	1036.44083	FT/MIN	AIR VELOCITY, DUCT 1, TYPICAL
		P4	.521246852	IN. W.G.	PRESSURE IN MAIN, FROM FAN 1
		K2	1.898097E-5		
		Q2	90.4006724	CFM	AIR FLOW FROM FAN 1
		V2	1036.44083		
		P3	.686121197		
		K3	2.609884E-5		
		Q3	91.3091161	CFM	AIR FLOW FROM FAN 2
		V3	1046.85610		
		P6	.457115735		
		K4	4.954505E-6		
		Q4	181.709788	CFM	AIR FLOW FROM FAN 1 + FAN 2
		V4	925.909750		
		P5	.638838917		
		K5	2.847146E-5		
		Q5	93.5295652	CFM	AIR FLOW FROM FAN 3
		V5	1072.31348		
	.1	P7	1.295794E-5	IN. W.G.	PRESSURE AT END OF DUCT RUN
		K6	275.239353		
		Q6	1402.49352	CFM	TOTAL AIR FLOW
		V6			
	5	L1			
	4	D1			
	8	L2			
	4	D2			
	11	L3			
	4	D3			
	13	L4			
	6	D4			
	12	L5			
	4	D5			
	34	L6			

Variables

Status	Input	Name	Output	Unit	Comment
	6	D6			
		A1	.0872222222		
		A2	.0872222222		
		A3	.0872222222		
		VP1	.0669705680		
		VP3	.0683233160		
		VP5	.0716866840		
		A4	.19625		
		A5	.0872222222		
		A6	.19625		
		VP2	.0669705680		
		VP4	.0534481000		
		VP6	.1226299880		
	102.989569	F1			FAN CURVE COEFFICIENT FOR GREENHECK CSP-B110
	-40.3168557	F2			FAN CURVE COEFFICIENT FOR GREENHECK CSP-B110
	168.57961	F3			FAN CURVE COEFFICIENT FOR GREENHECK CSP-B110
	-525.308659	F4			FAN CURVE COEFFICIENT FOR GREENHECK CSP-B110
	877.553367	F5			FAN CURVE COEFFICIENT FOR GREENHECK CSP-B110
	-579.951975	F6			FAN CURVE COEFFICIENT FOR GREENHECK CSP-B110

11.03 parallel fans CSP B110.tkw

CHAPTER 12 – COMPRESSED AIR IN NETWORKS

FIG. 12.1 AIR TOOL SUPPLY

Compressed air is used extensively in industry and commercial applications such as for air tools.

The Darct-Weisbach equation is really not suitable for compressed air, so we will use the Harris Formula. From PIPING HANDBOOK, 5th Edition,

$DP = L*Q^2/(2390*Pc*D^{5.31})$

DP = Pressure drop, psi
L = Pipe length, feet
Q = Free air, cu.ft/min
Pc = Average line pressure, psi
D = Pipe id, inches

Now, for pressure drop between 2 points with a flow rate of Q1, pipe length L1 and pipe id of D1, DP = P1-P2 and, Pc = (P1+P2)/2

Then, $DP = L*Q^2/(2390*((P1+P2)/2)*D^{5.31}) = 2*L*Q^2/(2390*(P1+P2)*D^{5.31})$

Let $K1 = 2*L1/(2390*D1^{5.31})$, then $P1-P2 = K1*Q1^2/(P1+P2)$

PROBLEM 12.1 – AIR TOOL SUPPLY

This is a typical problem for a shop with air tools supplied by a central compressor. Several tools are used, all assumed to operate at 90 psi. The k factor (orifice factor) is calculated to provide a realistic usage. For instance, for an air hammer, SCFM=22 = k*(90)^0.5. Then k = 22/(90^.5) = 2.32. The tool symbols and k-factors will be used in the network diagram.

SYMBOL	TOOL	SCFM	k-FACTOR
A	AIR HAMMER	22	2.32
B	12" CIRC. SAW	17	1.79
C	7" DISC GRINDER	8	0.84
D	GREASE GUN	4	0.42
E	FLOOR JACK	6	0.63
F	½" IMPACT WRENCH	4	0.42
G	¾" IMPACT WRENCH	5	0.527
H	1" IMPACT WRENCH	10	1.054
I	PAINT SPRAYER	20	2.08

The objective is to find out the minimum discharge pressure at the compressor with all tools operating, given pipe sizes and lengths.

Discussion:

All pipe sizes were set at 1". The compressor was input with 92 psi outlet, the outlet pressures at each tool are close to 90. These outlet pressures can be adjusted by changing the compressor outlet pressure and/or pipe sizes. That can be left to the reader.

Discussion:

All pipe sizes were set at 1". The compressor was input with 92 psi outlet, the outlet pressures at each tool are close to 90. These outlet pressures can be adjusted by changing the compressor outlet pressure and/or pipe sizes. That can be left to the reader.

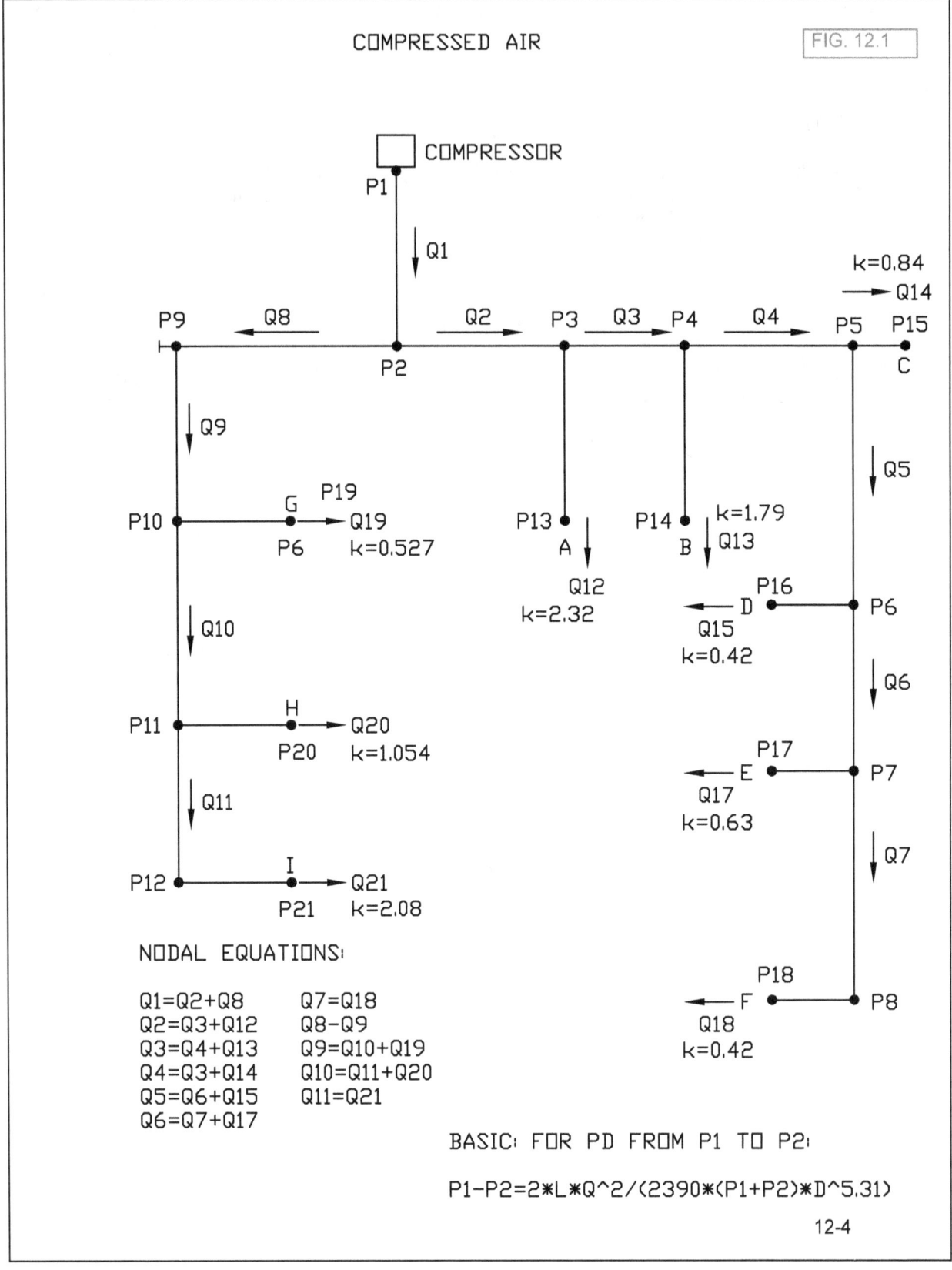

Rule

$P1-P2 = 2*L1*Q1^2/(2390*(P1+P2)*D1^{5.31})$
$P2-P3 = 2*L2*Q2^2/(2390*(P2+P3)*D2^{5.31})$
$P3-P4 = 2*L3*Q3^2/(2390*(P4+P3)*D3^{5.31})$
$P4-P5 = 2*L4*Q4^2/(2390*(P4+P5)*D4^{5.31})$
$P5-P15 = 2*L14*Q14^2/(2390*(P5+P15)*D14^{5.31})$
$P3-P13 = 2*L12*Q12^2/(2390*(P3+P13)*D12^{5.31})$
$P14-P4 = 2*L13*Q13^2/(2390*(P14+P4)*D13^{5.31})$

Q2=Q3+Q12
Q3=Q4+Q13

$P13=(Q12/2.32)^2$
$P14=(Q13/1.79)^2$
$P15=(Q14/0.84)^2$

$P5-P6 = 2*L5*Q5^2/(2390*(P5+P6)*D5^{5.51})$
$P6-P7 = 2*L6*Q6^2/(2390*(P6+P7)*D6^{5.31})$
$P7-P8 = 2*L7*Q7^2/(2390*(P7+P8)*D7^{5.31})$
$P6-P16 = 2*L16*Q15^2/(2390*(P6+P16)*D16^{5.31})$
$P7-P17 = 2*L17*Q17^2/(2390*(P7+P17)*D17^{5.31})$
$P8-P18 = 2*L18*Q18^2/(2390*(P8+P18)*D18^{5.31})$

$P16=(Q15/0.42)^2$
$P17=(Q17/0.63)^2$
$P18=(Q18/0.42)^2$

Q4=Q14+Q5
Q5=Q15+Q6
Q6=Q17+Q7
Q7=Q18

$P2-P9 = 2*L8*Q8^2/(2390*(P2+P9)*D8^{5.51})$
$P9-P10 = 2*L9*Q9^2/(2390*(P9+P10)*D9^{5.51})$
$P10-P11 = 2*L10*Q10^2/(2390*(P10+P11)*D10^{5.51})$
$P11-P12 = 2*L11*Q11^2/(2390*(P11+P12)*D11^{5.51})$
$P10-P19 = 2*L19*Q19^2/(2390*(P19+P10)*D19^{5.51})$
$P11-P20 = 2*L20*Q20^2/(2390*(P11+P20)*D20^{5.51})$
$P12-P21 = 2*L21*Q21^2/(2390*(P12+P21)*D21^{5.51})$

$P19=(Q19/0.527)^2$
$P20=(Q20/1.054)^2$
$P21=(Q21/2.08)^2$

Q1=Q2+Q8
Q8=Q9
Q9=Q19+Q10
Q10=Q20+Q11
Q11=Q21

Variables

Status	Input	Name	Output	Unit	Comment
	92	P1		PSI	OUTPUT PRESSURE AT COMPRESSOR
Guess	1	P2		PSI	
	87	L1		FEET	PIPE LENGTH 1, TYPICAL
Guess	1	Q1		CFM	TOTAL AIR FLOW
Guess	1	P3		PSI	
Guess	1	D1			
	45	L2			
Guess	1	Q2		CFM	
Guess	1	D2			
		P4		PSI	
	20	L3			
Guess	1	Q3		CFM	
Guess	1	D3			
Guess	1	P5		PSI	
	40	L4			
Guess	1	Q4		CFM	
Guess	1	D4			
Guess	1	P15		PSI	OUTLET PRESSURE
	15	L14			
Guess	1	Q14		CFM	OUTLET FLOW
Guess	1	D14			
Guess	1	P13		PSI	OUTLET PRESSURE
	30	L12			
Guess	1	Q12		CFM	OUTLET FLOW
Guess	1	D12			
Guess	1	P14		PSI	OUTLET PRESSURE
	30	L13			
Guess	1	Q13		CFM	OUTLET FLOW
Guess	1	D13			
Guess	1	P6		PSI	OUTLET PRESSURE
	15	L5			
Guess	1	Q5		CFM	OUTLET FLOW
Guess	1	D5			
Guess	1	P7		PSI	OUTLET PRESSURE
	25	L6			
Guess	1	Q6		CFM	OUTLET FLOW
Guess	1	D6			

Variables

Status	Input	Name	Output	Unit	Comment
Guess	1	P8		PSI	OUTLET PRESSURE
Guess	16	L7			
Guess	1	Q7		CFM	OUTLET FLOW
Guess	1	D7			
Guess	1	P16		PSI	OUTLET PRESSURE
Guess	15	L16			
Guess	1	Q15		CFM	OUTLET FLOW
Guess	1	D16			
Guess	1	P17		PSI	OUTLET PRESSURE
Guess	22	L17			
Guess	1	Q17		CFM	OUTLET FLOW
Guess	15	D17			
Guess	1	P18		PSI	OUTLET PRESSURE
Guess	15	L18			
Guess	1	Q18		CFM	OUTLET FLOW
Guess	1	D18			
Guess	1	P9		PSI	
Guess	34	L8			
Guess	1	Q8		CFM	
Guess	1	D8			
Guess	1	P10		PSI	
Guess	20	L9			
Guess	1	Q9		CFM	
Guess	1	D9			
Guess	1	P11		PSI	
Guess	15	L10			
Guess	1	Q10		CFM	
Guess	1	D10			
Guess	1	P12		PSI	
Guess	15	L11			
Guess	1	Q11		CFM	
Guess	1	D11			
Guess	1	P19		PSI	OUTLET PRESSURE
Guess	15	L19			
Guess	1	Q19		CFM	OUTLET FLOW
Guess	1	D19			
Guess	1	P20		PSI	OUTLET PRESSURE

Variables

Status	Input	Name	Output	Unit	Comment
	15	L20			
Guess	1	Q20		CFM	OUTLET FLOW
Guess	1	D20			
	1	P21		PSI	OUTLET PRESSURE
	15	L21			
Guess	1	Q21		CFM	OUTLET FLOW
Guess	1	D21			

ca part 1 and 2. and 3tkw RULES.tkw

Status	Input	Name	Output	Unit	Comment
	92	P1		PSI	OUTPUT PRESSURE AT COMPRESSOR
		P2	88.40397338	PSI	
	87	L1		FEET	PIPE LENGTH 1, TYPICAL
		Q1	94.39713747	CFM	TOTAL AIR FLOW
		P3	87.6325747	PSI	
	1	D1			
	45	L2			
		Q2	60.05076747	CFM	
	1	D2			
		P4	87.49208526	PSI	
	20	L3			
		Q3	38.3410906	CFM	
	1	D3			
		P5	87.40284016	PSI	
	40	L4			
		Q4	21.59408236	CFM	
	1	D4			
		P15	87.3984118!	PSI	OUTLET PRESSURE
	15	L14			
		Q14	7.85291788!	CFM	OUTLET FLOW
	1	D14			
		P13	87.5650392	PSI	OUTLET PRESSURE
	30	L12			
		Q12	21.7096768	CFM	OUTLET FLOW
	1	D12			
		P14	87.5323133!	PSI	OUTLET PRESSURE
	30	L13			
		Q13	16.7470082	CFM	OUTLET FLOW
	1	D13			
		P6	87.3892805	PSI	OUTLET PRESSURE
	15	L5			
		Q5	13.7411644	CFM	OUTLET FLOW
	1	D5			
		P7	87.3777489(	PSI	OUTLET PRESSURE
	25	L6			
		Q6	9.81493551	CFM	OUTLET FLOW
	1	D6			

Variables   ca part 1 and 2. and 3tkw RULES.tkw

Status	Input	Name	Output	Unit	Comment
	16	P8	87.37656800	PSI	OUTLET PRESSURE
		L7			
	1	Q7	3.92594336	CFM	OUTLET FLOW
		D7			
	15	P16	87.3881734	PSI	OUTLET PRESSURE
		L16			
	1	Q15	3.92622895	CFM	OUTLET FLOW
		D16			
	22	P17	87.3777489	PSI	OUTLET PRESSURE
		L17			
	15	Q17	5.88899215	CFM	OUTLET FLOW
		D17			
	15	P18	87.3754609	PSI	OUTLET PRESSURE
		L18			
	1	Q18	3.92594336	CFM	OUTLET FLOW
		D18			
	34	P9	88.2139364	PSI	
		L8			
	1	Q8	34.3463700	CFM	
		D8			
	20	P10	88.1019585	PSI	
		L9			
	1	Q9	34.3463700	CFM	
		D9			
	15	P11	88.0403629	PSI	
		L10			
	1	Q10	29.3998576	CFM	
		D10			
	15	P12	88.0132223	PSI	
		L11			
	1	Q11	19.5105855	CFM	
		D11			
	15	P19	88.1002155	PSI	OUTLET PRESSURE
		L19			
	1	Q19	4.94651238	CFM	OUTLET FLOW
		D19			
		P20	88.0333909	PSI	OUTLET PRESSURE

Variables

ca part 1 and 2. and 3tkw RULES.tkw

Status	Input	Name	Output	Unit	Comment
	15	L20			
		Q20	9.88927209	CFM	OUTLET FLOW
	1	D20			
		P21	87.9860733	PSI	OUTLET PRESSURE
	15	L21			
		Q21	19.5105855	CFM	OUTLET FLOW
	1	D21			

CHAPTER 13 – NATURAL GAS

Natural gas flow is calculated from the equations in the International Fuel Gas Code (IFGC) as follows: **$D = Q^{0.381}/(19.17*((Dh/(Cr*L))^{0.206})$ (for Low pressure - < 1.5 psi)**

Where:

D = inside diameter of pipe, inches
Q = input rate applicance(s), cu.ft/hour
Dh = Upstream pressure, psia – Downstream pressure, psia (P1-P2)
Cr = 0.6094 (for natural gas)
L = Equivalent length of pipe, feet

For practicality, use the charts in the IFGC for sizing.

If however, you want to size each line to the minimum, this procedure will work. The

Set the minimum pressure at each outlet, " w.g. Knowing the minimum input pressure, maximum pressure drop = Input pressure – minimum outlet pressure. The rate of pressure drop = Total length/(Input pressure – minimum outlet pressure). Total up the length to the furtherest outlet. This is called the total developed length (TDL).

Let maximum supply pressure = Pmax and minimum supply pressure at furtherest outlet = Pmin. Let total developed length = TDL. Let RATE = (Pmax-Pmin)/TDL. Set each pressure drop between pipe lengths (where flow rate changes) = Dp/Length = RATE. With this criteria, each pipe length will have a maximum allowable pressure drop to minimize the pipe size. The above equation (from the code) for D is used.

The example problem uses a maximum supply pressure = 1.4psi = 38.78 "w,g, Set minimum pressure at furtherest outlet = 20"w.g. Then Pmax-Pmin = 38.78-20 = 18.78. Then, rate = 18.78/TDL. This ensures the furtherest outlet receives 20" w.g. pressure, using the smallest pipe size. See problem 13.01. The minimum sizes are sized for. Of course, the next available commercial pipe size will be chosen.

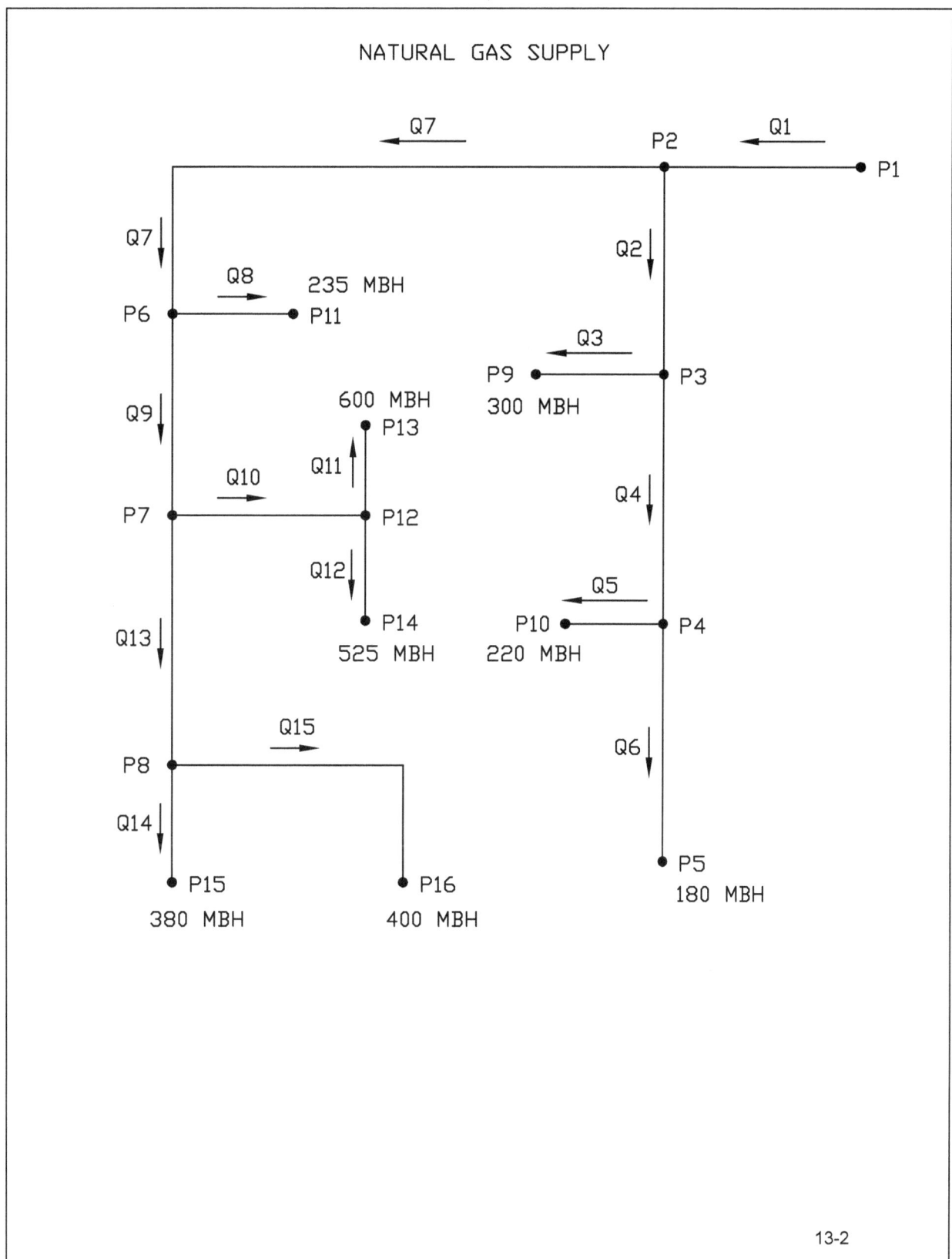

Rule

```
D1=Q1^0.381/(19.17*((P1-P2)/(0.6094*L1))^0.206)
D2=Q2^0.381/(19.17*((P2-P3)/(0.6094*L2))^0.206)
D3=Q3^0.381/(19.17*((P3-P9)/(0.6094*L3))^0.206)
D4=Q4^0.381/(19.17*((P3-P4)/(0.6094*L4))^0.206)
D5=Q5^0.381/(19.17*((P4-P10)/(0.6094*L5))^0.206)
D6=Q6^0.381/(19.17*((P4-P5)/(0.6094*L6))^0.206)
D7=Q7^0.381/(19.17*((P2-P6)/(0.6094*L7))^0.206)
D8=Q8^0.381/(19.17*((P6-P11)/(0.6094*L8))^0.206)
D9=Q9^0.381/(19.17*((P6-P7)/(0.6094*L9))^0.206)
D10=Q10^0.381/(19.17*((P7-P12)/(0.6094*L10))^0.206)
D11=Q11^0.381/(19.17*((P12-P13)/(0.6094*L11))^0.206)
D12=Q12^0.381/(19.17*((P12-P14)/(0.6094*L12))^0.206)
D13=Q13^0.381/(19.17*((P7-P8)/(0.6094*L13))^0.206)
D14=Q14^0.381/(19.17*((P8-P15)/(0.6094*L14))^0.206)
D15=Q15^0.381/(19.17*((P8-P16)/(0.6094*L15))^0.206)

;>>>/* SET PRESSURE DROP RATE */
TDL=L1+L7+L9+L13+L15
TPD=18.78
RATE=TPD/TDL

;>>>/* INPUT FLOW RATES */
Q1=2840
Q2=700
Q3=300
Q4=400
Q5=220
Q6=180
Q7=2140
Q8=235
Q9=1905
Q10=1125
Q11=600
Q12=525
Q13=780
Q14=380
Q15=400

;>>>/* INPUT PIPE LENGTHS */
L1=340
L2=69
L3=77
L4=24
L5=87
L6=45
L7=180
L8=35
L9=56
L10=87
```

Rule
L11=14
L12=21
L13=22
L14=55
L15=78

;>>>/* SET P1 */
P1=38.78
(P1-P2)/L1=RATE
(P2-P3)/L2=RATE
(P3-P9)/L3=RATE
(P3-P4)/L4=RATE
(P4-P10)/L5=RATE
(P4-P5)/L6=RATE
(P2-P6)/L7=RATE
(P6-P11)/L8=RATE
(P6-P7)/L9=RATE
(P7-P12)/L10=RATE
(P12-P13)/L11=RATE
(P12-P14)/L12=RATE
(P7-P8)/L13=RATE
(P8-P15)/L14=RATE
(P8-P16)/L15=RATE

Status	Input	Name	Output	Unit	Comment
Guess	1	D1		INCHES	CALCULATED DIAMETER, TYPICAL
		Q1		CU.FT/HR	FLOW RATE, GIVEN
		P1		"W.G.	INLET PRESSURE, GIVEN
		P2		"W.G.	
		L1		FEET	LENGTH OF PIPE 1, TYPICAL
Guess	1	D3		INCHES	CALCULATED MINIMUM PIPE SIZE
		Q3		CU.FT/HR	
		P3		"W.G.	
		P9		"W.G.	
		L3			
Guess	1	D4		INCHES	CALCULATED MINIMUM PIPE SIZE
		Q4		CU.FT/HR	
		P4		"W.G.	
		L4			
Guess	1	D5		INCHES	CALCULATED MINIMUM PIPE SIZE
		Q5		CU.FT/HR	
		P10		"W.G.	
		L5			
Guess	1	D6		INCHES	CALCULATED MINIMUM PIPE SIZE
		Q6		CU.FT/HR	
		P5		"W.G.	
		L6			
Guess	1	D7		INCHES	CALCULATED MINIMUM PIPE SIZE
		Q7		CU.FT/HR	
		P6		"W.G.	
		L7			
Guess	1	D8		INCHES	CALCULATED MINIMUM PIPE SIZE
		Q8		CU.FT/HR	
		P11		"W.G.	
		L8			
Guess	1	D9		INCHES	CALCULATED MINIMUM PIPE SIZE
		Q9		CU.FT/HR	
		P7		"W.G.	
		L9			
Guess	1	D10		INCHES	CALCULATED MINIMUM PIPE SIZE
		Q10		CU.FT/HR	
		P12		"W.G.	

Variables
LP GAS PIPE SIZES.tkw

Status	Input	Name	Output	Unit	Comment
Guess	1	L10			
		D11		INCHES	CALCULATED MINIMUM PIPE SIZE
		Q11		CU.FT/HR	
		P13		"W.G.	
		L11			
Guess	1	D12		INCHES	CALCULATED MINIMUM PIPE SIZE
		Q12		CU.FT/HR	FLOW RATE TO APPLIANCE
		P14		"W.G.	
		L12			
Guess	1	D13		INCHES	CALCULATED MINIMUM PIPE SIZE
		Q13		CU.FT/HR	FLOW RATE TO APPLIANCE
		L13			
Guess	1	D14		INCHES	CALCULATED MINIMUM PIPE SIZE
		Q14		CU.FT/HR	FLOW RATE TO APPLIANCE
		P8		"W.G.	
		P15		"W.G.	
		L14			
Guess	1	D15		INCHES	CALCULATED MINIMUM PIPE SIZE
		Q15		CU.FT/HR	FLOW RATE TO APPLIANCE
		P16		"W.G.	MINIMUM PRESSURE AT FARTHEREST USER
		L15			
		TDL		FEET	TOTAL DEVELOPED LENGTH, TO FARTHEREST USER
		TPD		"W.G.	TOTAL PRESSURE DROP ALLOWED
		RATE		"W.G./FT	PRESSURE DROP RATE
		Q2		CU.FT/HR	
		L2			
Guess	1	D2		INCHES	CALCULATED MINIMUM PIPE SIZE

Variables   LP GAS PIPE SIZES.tkw

Status	Input	Name	Output	Unit	Comment
		D1	2.038777381	INCHES	CALCULATED DIAMETER, TYPICAL
		Q1	22840	CU.FT/HR	FLOW RATE, GIVEN
		P1	38.78	"W.G.	INLET PRESSURE, GIVEN
		P2	29.3344378	"W.G.	
		L1	340	FEET	LENGTH OF PIPE 1, TYPICAL
		D3	.865842138	INCHES	CALCULATED MINIMUM PIPE SIZE
		Q3	300	CU.FT/HR	
		P3	27.4175443	"W.G.	
		P9	25.2784023	"W.G.	
		L3	77		
		D4	.966140697	INCHES	CALCULATED MINIMUM PIPE SIZE
		Q4	400	CU.FT/HR	
		P4	26.7507988	"W.G.	
		L4	24		
		D5	.769340439	INCHES	CALCULATED MINIMUM PIPE SIZE
		Q5	220	CU.FT/HR	
		P10	24.3338461	"W.G.	
		L5	87		
		D6	.712712448	INCHES	CALCULATED MINIMUM PIPE SIZE
		Q6	180	CU.FT/HR	
		P5	25.5006508	"W.G.	
		L6	45		
		D7	1.830388803	INCHES	CALCULATED MINIMUM PIPE SIZE
		Q7	2140	CU.FT/HR	
		P6	24.3338461	"W.G.	
		L7	180		
		D8	.788918927	INCHES	CALCULATED MINIMUM PIPE SIZE
		Q8	235	CU.FT/HR	
		P11	23.3615088	"W.G.	
		L8	35		
		D9	1.751037744	INCHES	CALCULATED MINIMUM PIPE SIZE
		Q9	1905	CU.FT/HR	
		P7	22.7781065	"W.G.	
		L9	56		
		D10	1.432664934	INCHES	CALCULATED MINIMUM PIPE SIZE
		Q10	1125	CU.FT/HR	
		P12	20.3611538	"W.G.	

Status	Input	Name	Output	Unit	Comment
		L10	87		
		D11	1.127537894	INCHES	CALCULATED MINIMUM PIPE SIZE
		Q11	600	CU.FT/HR	
		P13	19.9722189	"W.G.	
		L11	14		
		D12	1.07160866	INCHES	CALCULATED MINIMUM PIPE SIZE
		Q12	525	CU.FT/HR	FLOW RATE TO APPLIANCE
		P14	19.77775141	"W.G.	
		L12	21		
		D13	1.24607322	INCHES	CALCULATED MINIMUM PIPE SIZE
		Q13	780	CU.FT/HR	FLOW RATE TO APPLIANCE
		L13	22		
		D14	.947442954	INCHES	CALCULATED MINIMUM PIPE SIZE
		Q14	380	CU.FT/HR	FLOW RATE TO APPLIANCE
		P8	22.16692301	"W.G.	
		P15	20.6339645	"W.G.	
		L14	55		
		D15	.9661406971	INCHES	CALCULATED MINIMUM PIPE SIZE
		Q15	400	CU.FT/HR	FLOW RATE TO APPLIANCE
		P16	20	"W.G.	MINIMUM PRESSURE AT FARTHEREST USER
		L15	78		
		TDL	676	FEET	TOTAL DEVELOPED LENGTH, TO FARTHEREST USER
		TPD	18.78	"W.G.	TOTAL PRESSURE DROP ALLOWED
		RATE	.027781065	"W.G./FT	PRESSURE DROP RATE
		Q2	?700	CU.FT/HR	
		L2	69		
		D2	1.195743077	INCHES	CALCULATED MINIMUM PIPE SIZE

CHAPTER 14 – FINISH AND REFERENCES

This is the conclusion of this text. We covered a lot of problems, but there are many more out there. This is by no means the totality of piping network solutions, just one flavor. The techniques shown can be combined to solve other network problems. The reader is highly encouraged to study more and apply the lessons learned. The author would greatly appreciate feedback and how the book was useful or if there are questions.

I may be reached at tt802701@gmail.com

REFERENCES

1. Cameron Hydraulic Data by Ingersoll-Rand, 16th edition.
2. Crane's Flow of Fluids, Technical Paper 410, 19the printing.
3. Standard Handbook for Mechanical Engineers, 7th edition.
4. Schaum's 2500 Solved Problems in Fluid Mechanics and Hydraulics, 1st Edition.
5. Piping Handbook by Sabin Crocker, 1945 (oldie but goodie)..
6. Piping Handbook, 5th Edition.
7. Handbook of Chemistry and Physics, McGraw Hill, 35th Edition.
8. Air Distribution, by Carrier Air Conditioning, Inc.

Compute friction factors fast for flow in pipes

This survey shows how explicit equations for the friction factor during turbulent flow in pipes replace graphical methods.

Žarko Olujić, Zagreb University

☐ In engineering practice, friction factors are calculated for turbulent flow from the following:

1. *Hydraulically smooth pipes*—Prandtl's [1] universal law-of-friction equation:

$$1/\sqrt{f} = 2\log(N_{Re}\sqrt{f}) - 0.8 \quad (1)$$

2. *All pipes (hydraulically smooth and rough)*—the Colebrook and White [1,2] equation:

$$\frac{1}{\sqrt{f}} = -2\log\left(\frac{\epsilon/d}{3.7} + \frac{2.51}{N_{Re}\sqrt{f}}\right) \quad (2)$$

where f is the Darcy (sometimes called Moody) friction factor, which is four times the Fanning friction factor, ϵ/d is relative roughness, N_{Re} is the Reynolds number.

The widely used Moody chart [3] is based on Eq. (2), which is an interpolation formula. Its theoretical background can be found elsewhere [4]. For smooth pipes where the relative roughness, ϵ/d, is very small, the first term in the bracket of Eq. (2) can be neglected. Then, Eq. (1) and Eq. (2) become identical.

The limits for using Eq. (1) and (2) are generally taken to be (3,000 to 4,000) $\leq N_{Re} \leq 10^8$. Less-known experimental data obtained by Murin [5] and presented in part by Round [6] confirm the reliability of the Colebrook and White equation. According to Smislov [7], Murin investigated friction factors for turbulent flow of cold and hot water in brass pipes, and in 49 new, cleaned, and used commercial steel pipes (0.04 m $\leq d \leq$ 0.14 m) for different applications. The results of these investigations are summarized in Fig. 1.

The problem with Eq. (1) and (2) is in the implicitness of the friction factor, f. Thus, some sort of trial-and-error method for calculating f from given values of N_{Re} and ϵ/d is required. In order to overcome these difficulties, explicit equations have been proposed.

A simple explicit equation that can replace Eq. (1) within an accuracy of $\pm 1\%$ is a less-known approximation also proposed by Colebrook [1,2]:

$$f = (1.8\log(N_{Re}/7))^{-2} \quad (3)$$

Filonenko [8] proposed an empirical equation:

$$f = [1.82\log(N_{Re}) - 1.64]^{-2} \quad (4)$$

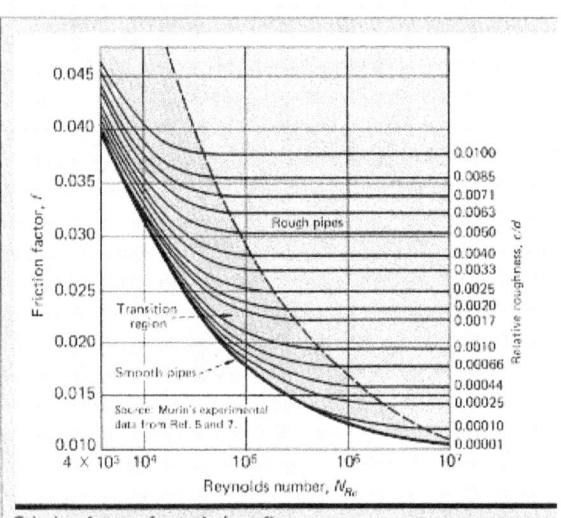

Friction factors for turbulent flow in commercial steel pipes   Fig. 1

Friction factors for turbulent flow in smooth pipes, as calculated from equations   Table I

	Friction factors calculated by equation of:			
Reynolds no., N_{Re}	Prandtl, Eq. (1)	Colebrook, Eq. (3)	Filonenko-Altshul, Eq. (4) and (5)	Konakov, Eq. (6)
4,000	0.0399	0.0406	0.0414	0.0403
10^4	0.0309	0.0310	0.0314	0.0308
$5\cdot10^4$	0.0209	0.0208	0.0209	0.0207
10^5	0.0180	0.0179	0.0180	0.0178
$2\cdot10^5$	0.0156	0.0155	0.0156	0.0155
$5\cdot10^5$	0.0132	0.0131	0.0131	0.0130
10^6	0.0116	0.0116	0.0116	0.0116
$2\cdot10^6$	0.0104	0.0104	0.0104	0.0103
$5\cdot10^6$	0.0090	0.0090	0.0090	0.0090
10^7	0.0081	0.0081	0.0081	0.0081
10^8	0.0059	0.0060	0.0060	0.0060

Comparison of calculated and experimental values for the friction factor for turbulent flow in smooth pipes — Table II

Reynolds no., N_{Re}	Source*	\multicolumn{6}{c}{Friction factors, f, for relative roughness, ϵ/d, of:}					
		0.05	10^{-2}	10^{-3}	10^{-4}	10^{-5}	10^{-6}
4,000	1	0.0769	0.0491	0.0409	0.0400	0.0399	0.0399
	2	0.0765	0.0506	0.0417	0.0407	0.0406	0.0406
	3	0.0769	0.0491	0.0408	0.0399	0.0398	0.0398
	4	0.0560	0.0446	0.0403	0.0398	0.0397	0.0397
	5	0.0655	0.0471	0.0413	0.0407	0.0406	0.0406
	6	0.0715	0.0484	0.0407	0.0398	0.0397	0.0397
	7	0.0769	0.0488	0.0406	0.0397	0.0396	0.0396
	8	0.0770	0.0491	0.0409	0.0400	0.0399	0.0399
	E		0.0462	0.0408	0.0408	0.0408	0.0408
10^4	1	0.0737	0.0431	0.0324	0.0310	0.0309	0.0309
	2	0.0750	0.0440	0.0327	0.0312	0.0310	0.0310
	3	0.0738	0.0432	0.0324	0.0310	0.0309	0.0309
	4	0.0537	0.0396	0.0327	0.0317	0.0316	0.0316
	5	0.0613	0.0402	0.0322	0.0311	0.0310	0.0310
	6	0.0680	0.0424	0.0320	0.0306	0.0304	0.0304
	7	0.0738	0.0431	0.0323	0.0310	0.0309	0.0309
	8	0.0738	0.0431	0.0324	0.0310	0.0309	0.0309
	E		0.0415	0.0315	0.0315	0.0315	0.0315
10^5	1	0.0717	0.0385	0.0222	0.0185	0.0180	0.0180
	2	0.0719	0.0387	0.0223	0.0185	0.0179	0.0179
	3	0.0717	0.0385	0.0222	0.0186	0.0181	0.0180
	4	0.0522	0.0354	0.0223	0.0184	0.0178	0.0178
	5	0.0586	0.0350	0.0217	0.0184	0.0179	0.0179
	6	0.0658	0.0380	0.0226	0.0183	0.0177	0.0176
	7	0.0718	0.0385	0.0222	0.0186	0.0182	0.0181
	8	0.0718	0.0385	0.0222	0.0185	0.0180	0.0180
	E		0.0378	0.0211	0.0185	0.0180	0.0176
10^6	1	0.0715	0.0379	0.0199	0.0134	0.0119	0.0117
	2	0.0716	0.0380	0.0200	0.0136	0.0119	0.0116
	3	0.0715	0.0379	0.0200	0.0135	0.0119	0.0117
	4	0.0520	0.0348	0.0199	0.0125	0.0103	0.0100
	5	0.0583	0.0344	0.0196	0.0136	0.0119	0.0116
	6	0.0655	0.0375	0.0208	0.0140	0.0118	0.0115
	7	0.0716	0.0380	0.0199	0.0135	0.0119	0.0118
	8	0.0716	0.0380	0.0199	0.0134	0.0119	0.0117
	E		0.0378	0.0197	0.0130	0.0121	
10^7	1	0.0715	0.0379	0.0197	0.0122	0.0090	0.0082
	2	0.0715	0.0379	0.0197	0.0122	0.0091	0.0083
	3	0.0715	0.0379	0.0197	0.0122	0.0090	0.0082
	4	0.0520	0.0348	0.0196	0.0112	0.0070	0.0058
	5	0.0583	0.0343	0.0193	0.0125	0.0093	0.0083
	6	0.0655	0.0375	0.0206	0.0131	0.0095	0.0083
	7	0.0716	0.0379	0.0197	0.0122	0.0090	0.0083
	8	0.0716	0.0379	0.0197	0.0122	0.0090	0.0082
	E		0.0378	0.0197	0.0120	0.0109	
10^8	1	0.0715	0.0379	0.0196	0.0120	0.0082	0.0064
	2	0.0715	0.0379	0.0196	0.0120	0.0082	0.0065
	3	0.0715	0.0379	0.0196	0.0120	0.0082	0.0064
	4	0.0520	0.0348	0.0196	0.0110	0.0063	0.0040
	5	0.0583	0.0343	0.0193	0.0124	0.0087	0.0067
	6	0.0655	0.0375	0.0206	0.0130	0.0090	0.0069
	7	0.0716	0.0379	0.0197	0.0122	0.0082	0.0064
	8	0.0716	0.0379	0.0196	0.0120	0.0080	0.0064
	E		0.0378	0.0197	0.0120		

*Source	Equation name	Equation no.	*Source	Equation name	Equation no.
1	Colebrook and White	(2)	6	Round	(9)
2	Churchill	(12)	7	Shacham	(10)
3	Chen	(11)	8	Shacham	(13)
4	Altshul	(8)	E	Murin's experimental data [6, 7]	...
5	Altshul	(7)			

Eq. (4) is almost identical to the Altshul [4] equation:

$$f = [1.82 \log(N_{Re}/100) + 2]^{-2} \quad (5)$$

Slightly different than the Filonenko-Altshul equation is the one proposed by Konakov [9]:

$$f = [1.8 \log(N_{Re}) - 1.5]^{-2} \quad (6)$$

Eq. (6) is almost identical to Eq. (3).

Table I shows the comparison for the values of f as calculated from Eq. (2), the Prandtl equation, and the several explicit equations given here. All of the explicit equations reproduce the Prandtl equation with remarkable accuracy in the range $4,000 \leq N_{Re} \leq 10^8$.

Equations for smooth and rough pipes

For predicting friction factors in all pipes, Altshul proposed two explicit equations:

$$f = \left\{1.82 \log\left[\frac{N_{Re}}{0.1 N_{Re}(\epsilon/d) + 7}\right]\right\}^{-2} \quad (7)$$

$$f = 0.11[(\epsilon/d) + (68/N_{Re})]^{0.25} \quad (8)$$

For smooth pipes, Eq. (8), widely used in Russian engineering practice [10], reduces to the well-known Blasius equation.

Recently, Round [6] slightly modified Eq. (7) in order to improve its accuracy at higher values of ϵ/d to:

$$f = \left\{1.8 \log\left[\frac{N_{Re}}{0.135 N_{Re}(\epsilon/d) + 6.5}\right]\right\}^{-2} \quad (9)$$

Substituting $f = 0.03$ in the right-hand side of Eq. (2), and applying the successive substitution twice, Shacham [11] derived the equation:

$$f = \left\{-2 \log\left[\frac{\epsilon/d}{3.7} - \frac{5.02}{N_{Re}} \log\left(\frac{\epsilon/d}{3.7} + \frac{14.5}{N_{Re}}\right)\right]\right\}^{-2} \quad (10)$$

Eq. (10) is similar to but much simpler than that proposed by Chen [12]:

$$f = \left[-2 \log\left(\frac{\epsilon/d}{3.7065} - Y\right)\right]^{-2} \quad (11a)$$

$$Y = \frac{5.0452}{N_{Re}} \log\left[\frac{(\epsilon/d)^{1.1098}}{2.8257} + Z\right] \quad (11b)$$

$$Z = 5.8506 (N_{Re})^{-0.8981} \quad (11c)$$

Churchill [13] proposed an equation that considers both the laminar- and transient-flow regimes:

$$f = 8\left[(8/N_{Re})^{12} + (A + B)^{-3/2}\right]^{1/12} \quad (12a)$$

$$A = \left[2.457 \ln\left(\frac{1}{(7/N_{Re})^{0.9} + 0.27(\epsilon/d)}\right)\right]^{16} \quad (12b)$$

$$B = (37,530/N_{Re})^{16} \quad (12c)$$

Shacham [11] also derived a more complicated equation than Eq. (10):

$$f = \left[\frac{X(1 - \ln X) - \frac{\epsilon/d}{3.7}}{1.15129 X + \frac{2.51}{N_{Re}}}\right]^{-2} \quad (13)$$

where X denotes the expression appearing inside the square brackets of Eq. (10).

Eq. (13) is more accurate than Eq. (10), which yields values for the friction factors within $\pm 1\%$ of those obtained from Eq. (2). As pointed out by Shacham, better accuracy in predicting friction factors is not needed.

Values for the friction factor, f, have been computed at a number of logarithmically equispaced values of the Reynolds number, N_{Re}, and the roughness factor, ϵ/d, for each of the explicit equations. These values are shown in Table II. Also included in Table II are friction factors computed from Eq. (2) and obtained from Fig. 1. Table II represents an extension of the results presented by Round [6].

The Colebrook and White relationship, Eq. (2), predicts friction factors that are in close agreement (within about $\pm 3\%$) with the experimental data of Murin. For the values computed from the explicit equations in Table II, we find that the Shacham, Chen and Churchill equations provide the best predictions relative to the Colebrook and White equation.

Conclusions

The explicit equations reproduce with high accuracy the implicit Prandtl equation for calculating friction factors for turbulent flow in hydraulically smooth pipes.

Among the equations proposed as alternatives to the implicit Colebrook and White equation, the Shacham equation [Eq. (10)] seems to be the most convenient one to use. Because of its accuracy and simplicity, Eq. (10) is recommended for practical use.

Steven Danatos, Editor

References

1. Richter, H., "Rohrhydraulik," 5th ed., pp. 116-168, Springer Verlag, Berlin/Heidelberg, 1971.
2. Colebrook, C. F., and White, C. M., *J. Inst. Civil Eng.*, Vol. 10, No. 1, pp. 99-118 (1937-1938).
3. Moody, L. F., *Trans. ASME*, Nov. 1944, pp. 671-684.
4. Altshul, A. D., and Kiselev, P. G., "Gidravlika i Aerodinamika" [Hydraulics and Aerodynamics], 2nd ed., pp. 166-196, Strojizdat, Moscow, 1975.
5. Murin, G. A., *Izv. Vses. Teplotekh. Inst.*, No. 10, pp. 21-27, 1948.
6. Round, G. F., *Can. J. Chem. Eng.*, Feb. 1980, pp. 157-176.
7. Smislov, V. V., "Gidravlika i Aerodinamika" [Hydraulics and Aerodynamics], pp. 157-176, Vysshaya Shkola, Kiev, 1979.
8. Filonenko, G. K., *Teploenergetika*, No. 4, pp. 15-21 (1954).
9. Konakov, V. K., *Dok. Akad. Nauk SSSR*, Vol. 25, No. 5, pp. 14-24 (1950).
10. Ideljchik, I. E., "Spravochnik po Gidravlicheskim Soprotivleniam" [Handbook of Hydraulic Resistances], 2nd ed., pp. 50-92, Mashinostroenie, Moscow, 1975.
11. Shacham, M., *Ind. Eng. Chem. Fundam.*, May 1980, pp. 228-229.
12. Chen, N. H., *Ind. Eng. Chem. Fundam.*, Aug. 1979, pp. 296-297.
13. Churchill, S. W., *Chem. Eng.*, Nov. 7, 1977, pp. 91-92.

The author

Zarko Olujić is assistant professor at the Institute of Chemical Engineering, Faculty of Technology, Zagreb University, 41000 Zagreb, Yugoslavia, where he does research and teaching in transport phenomena, unit operations and computer-aided design. He has performed experimental work on horizontal two-phase gas-liquid flow characteristics during his stay as research fellow at the University of Karlsruhe, W. Germany. He has a B.S. in petroleum technology and an M.S. and a D.Sc. in chemical engineering from Zagreb University.

www.ingramcontent.com/pod-product-compliance
Lightning Source LLC
Chambersburg PA
CBHW081719170526
45167CB00009B/3631